Science Networks. Historical Studies

 Birkhäuser

Science Networks. Historical Studies
Founded by Erwin Hiebert and Hans Wußing
Volume 54

Edited by Eberhard Knobloch, Helge Kragh and Volker Remmert

More information about this series at: http://www.springer.com/series/4883

Helge Kragh

Varying Gravity

Dirac's Legacy in Cosmology and Geophysics

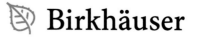 Birkhäuser

Helge Kragh
Niels Bohr Archive
Niels Bohr Institute
Copenhagen
Denmark

ISSN 1421-6329   ISSN 2296-6080   (electronic)
Science Networks. Historical Studies
ISBN 978-3-319-24377-1   ISBN 978-3-319-24379-5   (eBook)
DOI 10.1007/978-3-319-24379-5

Library of Congress Control Number: 2016931189

Springer Cham Heidelberg New York Dordrecht London

Cover illustration: From Waller Ms de-00215, August Beer: Über die Correction des Cosinusgesetzes bei
der Anwendung des Nicol'schen Prismas in der Photometrie, after 1850. With friendly permission by
The Waller Manuscript Collection (part of the Uppsala University Library Collections).

Printed on acid-free paper

Springer International Publishing AG Switzerland is part of Springer Science+Business Media
(www.birkhauser-science.com)

Preface

The reader will find here a historical investigation of a particular episode in the history of twentieth-century science which principally involves an unorthodox cosmological theory concerning the history of the universe and a no less unorthodox geological theory concerning the history of the Earth. By its very nature, the subject under examination, various early attempts of integrating cosmology and geophysics, is highly interdisciplinary. When Paul Dirac proposed that the gravitational constant decreases over cosmic time, a proposal which dates from the late 1930s, no one thought it would have consequences for geophysics. Nor did anyone think that the Earth might eventually be a testing ground for Dirac's hypothesis. After all, the domain and methods of geophysics were (and still are) very different from those of physical cosmology and at the time the two communities of scientists were entirely separate. As it happened, the audacious gravitation hypothesis was first applied to paleoclimatology in the late 1940s and about a decade later it entered geophysics as an argument for the expanding Earth.

The idea that the Earth has increased in size for at least 500 million years was at the time a fairly popular alternative to the resuscitated theory of continental drift that would soon be merged with mantle convection and sea floor spreading to develop into mainstream plate tectonics. The chief focus of the book is on the interconnection between the two hypotheses, but it also covers in some detail other aspects of varying gravity and the expanding Earth. The subjects gave rise to a considerable literature in physics, astronomy, cosmology, geology and geophysics, much of it of an interdisciplinary nature. Altogether several hundred scientific articles and a few books have been published on these subjects. However, from today's perspective, the efforts were wasted and may seem to have been just much ado about nothing. The currently established view is that the force of gravity, as given by the gravitational constant G, remains constant and that the radius of the Earth has not increased measurably since its formation some 4.5 billion years ago. In spite of this consensus view, there are still scientists cultivating either the varying-gravity hypothesis or the expanding Earth hypothesis—or, in a few cases, both hypotheses. But I largely keep to the historical ground, meaning the period up to about 1980, and only briefly refer to the modern scene.

I came to this subject initially as a result of my earlier studies of Dirac's physical theories and my work on the history of modern cosmology generally. Only at a later stage did I develop an interest in the history of the earth sciences in connection with courses in the history and philosophy of science given to undergraduate geology students at Aarhus University, Denmark. It was only then that I realized how relatively important the varying-gravity hypothesis and expanding Earth models were in the 1960s and 1970s. I recently published a couple of papers on the subject, one in *Physics in Perspective* and another in *History of Geo- and Space Sciences* (see the Bibliography). This book makes use of material from these papers but goes much beyond them. I should also mention that I am not the first to cover the subject. Paul Wesson examined it from a different and more scientific perspective in two valuable books dating from 1978 to 1980. However, Wesson primarily addressed his work to scientists and therefore paid little attention to the rich historical context of his subject.

The book is organized into four chapters of which the first one is rather brief and of an introductory nature, dealing essentially with developments before 1930. Chapter 2 investigates in detail the idea of varying gravity from a cosmological and physical perspective, starting with Dirac in 1937 and ending with the Jordan–Brans–Dicke gravitation theories of the early 1960s. While geophysics plays almost no role at all in this chapter, the expanding Earth is in the centre of Chap. 3 which deals in particular with theories that applied varying gravity as a mechanism for the assumed expansion of the Earth. The fourth chapter carries the story on until the early 1980s, at a time when varying-gravity hypotheses had proliferated but the expanding Earth hypothesis no longer enjoyed recognition from mainstream geophysicists. Although the approach of the book is neither biographical nor prosopographical, of course there are some scientists who appear more frequently than others. They include well-known physicists such as Paul Dirac, Pascual Jordan and Robert Dicke as well as the less well-known Hungarian geophysicist László Egyed. *Varying Gravity* ends with a rather lengthy bibliography which we hope can be useful to historians and scientists who might wish to explore further aspects of this case study.

Copenhagen, Denmark   Helge Kragh
August 2015

Contents

List of Figures

List of Tables

Chapter 1
Introductory Issues

Cosmology, the science of the universe at large, is of course very different from geology, a science that in its traditional meaning deals with only a single object in the vast universe, a planet called Earth. Yet the two sciences have interesting and often surprising interconnections that today are cultivated by a growing number of researchers. Both sciences—cosmology and the earth sciences—have changed drastically since the days of the scientific revolution in the seventeenth century, each in its own way and at different paces. In order to evaluate the events that occurred in the post-World War II period it will be useful to survey some of the earlier developments in a broad historical perspective. The survey in this chapter is meant to be an introduction only. It covers various attempts in the period up to the mid-1930s to think about the Earth in a cosmological perspective or otherwise to establish bridges between the science of the universe and that of the Earth. Until varying gravity entered the picture the two sciences had in common only the chronological problem, namely, the age of the Earth as related to the age of the universe.

1.1 The Heavens and the Earth

In the late seventeenth century, geology was often considered in what at the time was seen as a cosmological perspective, such as illustrated by the popularity of speculative "cosmogonies" principally dealing with the origin of the Earth. René Descartes' influential *Principia Philosophiae* from 1644 included a mechanical theory of the origin and evolution of the Earth which he derived from his general vortex theory of the universe. To him, cosmogony (in the sense of geogony, to use an antiquated term) was scarcely distinct from astronomy and cosmology. Other examples of this early tradition include classics such as Thomas Burnet's *The Sacred Theory of the Earth* from 1684, William Whiston's *New Theory of the*

© Springer International Publishing Switzerland 2016
H. Kragh, *Varying Gravity*, Science Networks. Historical Studies 54,
DOI 10.1007/978-3-319-24379-5_1

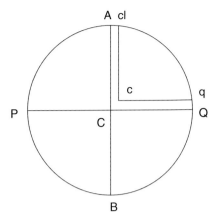

Fig. 1.1 Newton on the shape of the Earth (*Principia*). According to Newton, the proportion of the gravitational force in A from a sphere with radius CA to that in A from the actual Earth would be given by the ratio 126/125.5. From this and a few other arguments he concluded that the Earth was flattened by the ratio 1/230

Earth from 1696, and Leibniz's *Protogaea* published posthumously in 1749.[1] These and other works from the period were as much about God and the Bible as they were about the Earth. They were early examples of the powerful intellectual tradition known as physico-theology or "natural theology" that would dominate much scientific thinking until the beginning of the nineteenth century.

From the same period we have the first example ever of a fundamental physical theory changing the picture of the Earth, namely, Newton's successful prediction of the shape of the Earth based on his new theory of gravitation. Since the theory of gravitation was also the basis for Newtonian cosmology, the case can with some justification be seen as an early instance of the connection between cosmology and the earth sciences. In Book III of his *Philosophiae Naturalis Principia Mathematica* from 1687 Newton argued from mechanical principles that the rotating Earth must have the form of a slightly flattened oblate spheroid (Fig. 1.1). He estimated that the equatorial radius must exceed the polar radius by 1/229th part of the polar radius. The claim was controversial because the shape of the Earth was expected to be a prolate spheroid on the basis of Cartesian natural philosophy. Through the efforts of Pierre-Louis Maupertuis, Alexis-Claude Clairaut and other French scientists in the 1730s, it appeared that geodesic measurements agreed with Newton's prediction at least in a qualitative sense: the Earth was indeed slightly flattened at the poles.[2] The currently known flatness of the Earth is approximately 1 to 300, not far from Newton's value.

In *Principia* Newton also arrived at another result of great geophysical interest, the first scientifically based estimate of the mean density of the Earth. He reasoned as follows:

> In whatever way the planets were formed, at the time when the mass was fluid, all heavier matter made for the centre, away from the water. Accordingly, since the ordinary matter of our earth at its surface is about twice as heavy as water, and a little lower down, in mines, is

[1] See, for example, Laudan (1987) and Oldroyd (1996).

[2] The case is detailed in Greenberg (1995) and Terrall (2002).

found to be about three or four or even six times heavier than water, it is likely that the total amount of matter in the earth is about five to six times greater than it would be if the whole earth consisted of water.[3]

Although Newton's estimate of a density between 5 and 6 g cm^{-3} was little more than an educated guess, it happened to be correct.

During the next two centuries, geology was transformed into an arch-empirical, inductive science dealing in a much more narrow way than previously with the surface of the Earth and the natural causes for its changes. Yet the cosmological aspects did not vanish completely. What is often regarded as a watershed in the development of geology into a proper science, namely, the uniformitarian method-ology introduced by James Hutton and Charles Lyell between the 1790s and the 1830s, was in part of a cosmological nature. According to the uniformitarian credo, geological processes in the past were of the same kind as those observed today and they occurred at about the same rate and intensity as we observe. In his efforts to turn geology into a respectable science, Lyell insisted in his famous *Principles of Geology* (1830–1833) that the past history of the Earth can only be understood in terms of what is presently observed. Lyell explicitly based uniformitarianism on the belief that the laws of nature cannot possibly change in time or space. In the first volume of *Principles* from 1830 he rooted the new geology in "the permanency of the laws of nature," postulating that "their immutable constancy alone can enable us ... to arrive at the knowledge of general principles in the economy of our terrestrial system." A century later a few physicists would question the immutable constancy of Newton's law of gravitation, the very paradigm of natural law.

Lyell stressed that the new scientific geology was intimately related to the physical sciences but also, and no less importantly, that it was a science of its own which could not be subordinated either physics or other sciences. First and foremost, geology was entirely distinct from the traditional cosmogony dealing with the origin of the world. As Lyell wrote in the very beginning of *Principles*:

> The identification of its [geology's] objects with those of Cosmogony has been the most common and serious source of confusion. The first who endeavoured to draw a clear line of demarcation between these distinct departments, was Hutton, who declared that geology was in no ways concerned "with questions as to the origin of things." ... We shall attempt in the sequel of this work to demonstrate that geology differs as widely from cosmology, as speculations concerning the creation of man differ from history.[4]

Indeed, strict uniformitarianism was non-directional with respect to the arrow of time and for this reason alone it ruled out cosmogony. On the other hand, it did not follow that Lyell's geology was cosmologically irrelevant. For one thing, it implicitly entailed the notion of an eternal universe.

The uniformitarian principle would much later turn up in cosmological theory, in particular in the steady-state theory of the universe introduced in 1948. There is more than a superficial similarity between the geologists' uniformitarian principle

---

[3] Cohen (1999), p. 815.

[4] Lyell (1830), p. 4. Available online as http://www.esp.org/books/lyell/principles/facsimile/

and the "perfect cosmological principle" on which steady-state cosmology was founded.[5] The ordinary cosmological principle, which can be traced back long before the Einsteinian revolution in cosmology, states that on a very large scale the universe is spatially uniform, that is, homogeneous and isotropic. According to the perfect cosmological principle, a name due to Thomas Gold, the uniformity is also temporally valid: no era in the history of the universe is distinct from any other era. The perfect cosmological principle may seem to fit with the spirit of relativity theory's space-time symmetry, but what matters is that the principle disagrees with what we know about the universe. Perfect it may be, but wrong it is. In any case, for the large majority of geologists in the nineteenth century, Lyell included, the Earth was more than enough to deal with. They saw no reason to include other celestial bodies and even less to indulge into cosmological speculations.

Cosmology in the fin-de-siècle era remained foreign even to the majority of astronomers, who in the spirit of positivism preferred to ignore questions about the universe as a whole.[6] Works on "cosmogony" were not uncommon, but in most cases they dealt with the origin of the solar system or what the Dutch-British physicist Dirk ter Haar much later called "microcosmogony."[7] Other work from the period dealt with the mysterious spiral nebulae and their relation to the Milky Way. Indeed, terms such as "world" or "cosmos" typically referred to our solar system, usually discussed on the basis of Pierre Simon Laplace's popular nebular hypothesis or some modification of it. The material universe was often identified with the Milky Way, and "with the infinite possibilities beyond, science has no concern," as the Irish astronomer and historian of astronomy Agnes Mary Clerke summarily declared.[8]

Among the very few geologists in the Victorian period who adopted an astronomical perspective on the Earth was the versatile, self-educated Scotsman James Croll. Best known for his astronomical theory of terrestrial climate change, including the ice ages, Croll provided a rare link between the geologists and the astronomers. For example, his astronomically based theory directly influenced Lyell, the leading geologist of his day.[9] In 1885 Croll published a collection of essays under the intriguing title *Discussions on Climate and Cosmology*.[10] However, what he discussed was in reality the influence of astronomical causes on the climate of the Earth, and also subjects such as the age of the Earth, the nature and intensity of the Sun's heat, and the nature of the nebulae. He had nothing to say about cosmology in

---

[5] On the perfect cosmological principle and its history, see Balashov (1994) and Kragh (1996), pp. 182–183. See also Toulmin (1962) for an interesting case of a late-eighteenth-century geological author adopting a version of the perfect cosmological principle. More about the steady-state theory of the universe follows in Sect. 4.1.

[6] Merleau-Ponty (1983), Kragh (2008), pp. 152–157.

[7] ter Haar (1950), p. 132.

[8] Clerke (1890), p. 368. Of course, cosmology was not absent from pre-Einstein astronomy. For reviews, see North (1965) and Kragh (2007b).

[9] Fleming (2005).

[10] Croll (1885).

the wider sense of the term, the science of the universe in its totality. Indeed, the term "cosmology" only appeared in the book's title and not at all in the text.

Four years later Croll published another book in which he developed a cosmogonical impact theory that differed from the standard Laplace picture by assuming an initial state of dark bodies colliding at high velocities. His theory not only contradicted the nebular hypothesis but also the widely accepted Helmholtz–Kelvin gravitational theory of the origin of the Sun's heat. According to this theory the Sun was composed of compressible gases that through progressive contraction generated the heat and light that poured into space. "Are the facts of geology reconcilable with the theory?" Croll asked, referring to the theory of Kelvin (William Thomson) and Hermann von Helmholtz. He concluded that this was not the case.[11] At the end of his book he briefly considered the cosmological consequences of his own impact theory, which "on purely scientific grounds" led to an absolute beginning of the evolutionary universe. According to Croll, it followed that the universe would continue its evolution forever, whereas the thermal equilibrium state known as the "heat death" would never occur.

The heat death (*Wärmetod* in German) was perhaps the only truly cosmological phenomenon which entered both astronomy and the earth sciences in the period 1880–1920. The much-discussed problem was claimed to be a strict consequence of the second law of thermodynamics as stated in different versions by Kelvin and Rudolf Clausius in the mid-nineteenth century. According to the second law the entropy of the universe would continue to increase, eventually leading to an irreversible cessation of all physical processes—a dead universe including of course a dead Earth. Moreover, it apparently followed that the universe must have had a beginning in time corresponding to some minimum entropy.[12] Charles Darwin was among the naturalists of the Victorian era who worried about the relationship between biological evolution and the cosmic increase in entropy. He found it "intolerable" that the Sun and the planets would one day be doomed to annihilation.[13]

The twin problems of evolution and entropy were occasionally considered within the research tradition known as "cosmical physics," an interdisciplinary branch of science flourishing in the period from about 1890 to 1915 but with roots back to Alexander von Humboldt's influential *Kosmos* published in five volumes between 1845 and 1862. Cosmical physics, comprising elements of geophysics, meteorology and solar-terrestrial physics (including magnetic storms and the aurora borealis), was an ambitious attempt at synthetizing those parts of astronomy and the earth sciences that could be understood on the basis of the

---

[11] Croll (1889), p. 37.

[12] See Kragh (2008) for details and sources concerning the "entropic creation" argument.

[13] In his autobiography published in 1887, five years after his death, Darwin referred to "the view now held by most physicists, namely, that the sun with all the planets will in time grow too cold for life." He further reflected on "the mystery of the beginning of all things" but decided that it was "insoluble by us." Quoted in Kragh (2008), p. 108. The autobiography is available online at http://darwin-online.org.uk

laws of physics. A few of the cosmical physicists, including the Austrian meteorologist Wilhelm Trabert, the Swedish physical chemist Svante Arrhenius, and the Norwegian physicist Kristian Birkeland, adopted a cosmological perspective. For example, contrary to mainstream geophysics Birkeland maintained that magnetic storms and geomagnetism generally were influenced by magnetic disturbances in space, an idea which he related to his electromagnetic conception of the universe.[14]

In his massive *Lehrbuch der kosmischen Physik* (Textbook of Cosmical Physics) of 1903 as well as in his more popular *Worlds in the Making* of 1908, Arrhenius included chapters on cosmology and cosmogony. His own cosmological theory pictured an infinite stellar universe without a beginning or an end in time. The dispersive effects of the second law were kept in check by postulating an exchange of matter between stars and nebulae in the form of collisions, not unlike what Croll had suggested. The contractive effects of gravity on a cosmic scale were likewise kept in check by the expansive force of radiation pressure. Arrhenius' theory was to some extent inspired by geophysics in so far that it relied on the stellar radiation pressure that he inferred from the aurora borealis.

While Arrhenius was not a geologist, the American Thomas Chamberlin was professor of geology at the University of Chicago, the founder of the prestigious *Journal of Geology*, and the author (together with Rollin D. Salisbury) of a widely acclaimed textbook in geology. Together with his Chicago colleague, the astronomer Forest Ray Moulton, he developed a strong alternative to the nebular hypothesis of the formation of the Earth.[15] Interestingly, his dissatisfaction with the generally accepted nebular hypothesis was originally rooted in its geological rather than astronomical consequences. A specialist in glacial geology, Chamberlin was inspired to his "planetesimal theory" by considerations of the causes of glaciation and climatic change. In a book published in 1916 he related how his studies of glacial deposits in Wisconsin led him to the field of cosmogony. It occurred to him that the Laplace theory—"this theory of a simple decline from a fiery origin to a frigid end"—did not agree with the established record of glaciation in the past:

> When the inquiry was pressed still farther back, and support for the postulate of a molten globe was sought in the crust itself, it was not forthcoming. ... But one further step remained—to examine the cosmogonic postulates themselves. Could the earth ever have had the vast hot atmosphere postulated? Was the earth's gravity sufficient to hold so vast and vaporous an envelope at such high temperatures and in such an intense state of molecular activity as the old mode of genesis assigned? Was the gaseo-molten genesis a reality? Thus I was already across the pass that leads from the land of rocks into a realm of cosmogonic bogs and fens. ... Strangely enough, the cold trail of the ice invasion had led by this long and devious path into the nebulous field of genesis.[16]

---

[14] Birkeland's cosmological ideas, sometimes taken to be anticipations of modern "plasma cosmology," are dealt with in Kragh (2013).

[15] Brush (1996c), pp. 22–67, Fleming (2000).

[16] Quoted in MacMillan (1929), p. 4. As the Chicago astronomer William MacMillan remarked in his obituary of Chamberlin, the cosmogonical Chamberlin–Moulton hypothesis "furnished a foundation for the geologists, which is in harmony with the evidences of their own science."

The planetesimal theory that Chamberlin developed together with Moulton was for a period highly regarded not only by geologists but also by American astronomers. For example, in 1914 Vesto Melvin Slipher at the Lowell Observatory studied nebular spectra in order to test the Chamberlin–Moulton theory. This was only two years after he had detected the first nebular redshifts, a discovery that would have a revolutionary effect on cosmology.

1.2 Cosmology, Cosmogony, and Geology

Chamberlin was not the only professional geologist to consider connections between cosmology (or cosmogony) and the earth sciences. "The age of the earth may indeed claim an importance outside the geological science," wrote the prominent Irish geologist and physicist John Joly in a book of 1909. "For is it not the clue to the chronological scale of our solar system and—for want of better—even of our universe?"[17] To Joly, the common element that connected the Earth and the universe was radioactivity. The young British geologist and geophysicist Arthur Holmes published in 1913 the first edition of his classic *The Age of the Earth* in which he argued strongly for the use of radioactive methods in geochronology. Inspired by Arrhenius he broadened the scope to include also the time-scale of the universe as a whole. Estimating the oldest rocks to have an age about 1.7 billion years, he contrasted the finite age of the Earth with the supposedly infinite age of the universe. Holmes subscribed at the time to an eternal, ever-evolving cyclic universe of a kind similar to the one Arrhenius had advocated:

> If the development of the universe be everywhere towards equalisation of temperature implied by the laws of thermodynamics, the question arises—why in the abundance of time past, has this melancholy state not already overtaken us? . . . In the universe nothing is lost, and perhaps its perfect mechanism is the solitary and only possible example of perpetual motion. In its cyclic development we may find the secret of its eternity and discover that the dismal theory of thermal extinction is, after all, but a limited truth.[18]

The death of heat was ruled out by fiat. As we shall see in Sect. 3.5, this was not the last time that Holmes would use his geological expertise to deal with the universe at large.

The ideas of Arrhenius, Chamberlin, Joly, and Holmes were interesting but did little to strengthen the relationship between cosmology and the earth sciences except on a rhetorical level. In so far that a connection between the Earth and the heavens was admitted, still in the 1920s it was the more traditional one between geology and astronomy focusing on the Earth as a planet. At the time cosmology had changed dramatically as a result of new observations and, not least, the new models of the universe founded on Einstein's general theory of relativity. The

---

[17] Joly (1909), p. 212.

[18] Holmes (1913), p. 121.

famous astronomer Arthur Eddington was a specialist in Einstein's mathematically abstruse theory, including its cosmological aspects. Still, when he gave an address on the astronomy-geology relationship to the venerable Geological Society of London in 1922, he had nothing to say about cosmology in either its old or new meaning. Perhaps in consideration of his audience he chose to cover more traditional subjects such as the origin of the solar system, the age of the Earth, the Earth–Moon system, and the source of the Sun's heat.[19]

In his address Eddington also referred to the existence of radioactive minerals in the crust of the Earth. The presence of uranium and thorium, he said, indicated a "winding-up process [that] must have occurred under physical conditions vastly different from those in which we now observe only a running-down." Rather than locating the winding-up to a primordial state of the universe, he referred to "the intense heat in the interior of the stars." Eddington reasoned that the irreversible decay of radioactive bodies might function as an arrow of time in a sense similar to the entropic arrow of time. Arguments of this kind had been in the air for some time, but without attracting much interest among physicists, chemists and geologists.[20] It was only after the discovery of the expanding universe that the idea of a radioactive arrow of time became cosmologically significant.

To the Belgian pioneer cosmologist Georges Lemaître, a former student of Eddington, the idea suggested a universe of finite age that had come into existence in a radioactive inferno a period of time ago comparable to the half-lives of uranium and thorium. The argument from radioactive decay played an important role in the process that in 1931 led him to the hypothesis of an explosive universe originating in a "primeval atom," later referred to as the big-bang universe. As Lemaître recalled,

> The idea of this [primeval atom] hypothesis arose when it was noticed that natural radioactivity is a physical process which disappears gradually and which can, therefore, be expected to have been more important in earlier times. If it were not for a few elements of average lifetimes comparable to T_H [the Hubble time of the order 10^9 years], natural radioactivity would be completely extinct now. It might be thought, therefore, that radioactive elements did exist which are actually transformed into stable elements.[21]

The age of the Earth, a classic problem of geology, became of cosmological importance with the recognition in the early 1930s that most cosmological models based on the theory of general relativity resulted in an age of the universe less than two billion years.[22] For the class of so-called Friedmann models with zero

---

[19] Eddington (1923).

[20] See Kragh (2007a) for the early use of radioactive decay as a cosmic clock and generally the role of radioactivity in astrophysics and cosmology in the period between 1910 and the early 1930s.

[21] Lemaître (1949), p. 452.

[22] On the notorious time-scale problem, see Kragh (1996), pp. 73–79 and Brush (2001). The problem could be avoided if a positive cosmological constant was admitted, but this was a solution that few astronomers and physicists found appealing. In the period from about 1930 to the 1990s it was generally believed that $\Lambda = 0$.

cosmological constant and non-zero density value ($\Lambda = 0$, $\rho > 0$), the age t^* will always be smaller than the Hubble time T defined as the inverse of the Hubble constant H:

$$t^* = \alpha T = \alpha/H, \quad \alpha < 1.$$

Only for an empty universe will $\alpha = 1$, and even that does not help. The time-scale problem was that there were objects in the universe older than two billion years and hence older than the universe, which is obviously impossible. One of these objects is the Earth, which in the mid-1930s was known to be at least 3 billion years old. By comparison, with the accepted value of 1.8 billion years for the Hubble time, the age of the flat Einstein-de Sitter universe (given by 2/3 times the Hubble time) came out as 1.2 billion years. The time-scale problem did not rely crucially on the age of the Earth as determined by geologists and geochemists, for the stars and galaxies were thought to be much older still. Nonetheless, it was only the age of the Earth that was reliably determined with some degree of precision, and for this reason this geological datum held a special position in the cosmological discourse.[23] It would remain to do so until the 1950s, when it turned out that the Hubble time was much greater than what had been previously assumed. The presently accepted value, based on data from the Planck satellite, is $T = 13.82 \times 10^9$ years.

Although Einstein never looked for a connection between cosmological theory and geophysics, for a time he had an interest in problems of a geophysical nature. For example, between 1926 and 1933 he served on the editorial board of *Gerlands Beiträge zur Geophysik* edited by the Viennese climatologist Victor Conrad, a former professor of cosmical physics. Einstein associated with several Berlin geophysicists and meteorologists, including Adolf Schmidt, Julius Bartels and Heinrich Ficker. As early as 1919 he studied the changes in the Earth's moment of inertia caused by some of the partial tides of the Moon, and in 1926 he wrote a brief paper on geomorphology.[24] Near the end of his life Einstein corresponded with Charles Hapgood, an American historian and amateur geologist who developed an unorthodox theory of the displacements of the crust of the Earth.[25]

Among the geophysical problems that attracted Einstein's attention was the origin of geomagnetism and the possibility that it might be ascribed to the rotation of the Earth. At a meeting of the Swiss Physical Society in 1924, he suggested that the magnetic fields of the Earth and the Sun might be explained on the assumption

---

[23] Dehm (1949) is a useful review of the troubled situation at a time when the Hubble time was still believed to be of the order of 2 billion years.

[24] See Schröder and Treder (1997), according to whom "Albert Einstein initiated geophysical research with his works and contributed to studies that were often interdisciplinary in character."

[25] For the Einstein–Hapgood connection and other aspects of Einstein's interest in the earth sciences, see Martinez-Frias et al. (2006).

that the numerical charge of the proton q_p slightly exceeded that of the electron, q_e.[26] Specifically he suggested

$$|q_p/q_e| = 1 + \varepsilon, \quad \varepsilon = 3 \times 10^{-19}.$$

In favour of the hypothesis, Einstein pointed out the order-of-magnitude relationship

$$\varepsilon q_e \cong m\sqrt{G},$$

where m is the electron's mass. Einstein, who at the time tried to develop a unified field theory, believed that such a connection between fundamental electrical and gravitational quantities might be more than a mere coincidence. However, Einstein's charge-excess hypothesis was quickly shot down by two Swiss physicists (A. Piccard and E. Kessler) who found experimentally that $\varepsilon < 5 \times 10^{-21}$. Nonetheless, somewhat similar ideas of connections between gravity, electricity and geomagnetism were later pursued by several other physicists, including Patrick Blackett in England (see Sect. 2.3). Einstein did not publish his speculations concerning a "gravomagnetic effect" in 1924, but he referred to them in an address on the ether the same year and in 1928 he discussed the formula for the effect in a communication to the Prussian Academy of Sciences.

To summarize, although geological data and reasoning were in a few cases considered relevant to cosmology, or vice versa, before World War II the two areas of science were essentially separate and particularly so if the term "cosmology" is taken in its wide meaning as the science of the universe in its totality. By and large, geochronology was the only exception. By contrast, there was much interest in the astronomy-geology interface, especially as regards the ice ages and other aspects of the past climate of the Earth.[27] Among the celestial bodies, only the Moon and the Sun were considered relevant for geophysics.

However, one more area deserves to be mentioned as an early link between cosmology and the earth sciences, namely geochemistry. Even before the discovery of the atomic nucleus many scientists speculated that the atoms of the chemical elements were composite bodies that had evolved from a more primitive substance in the cosmic past. One of the scientists was Croll, who dealt with the subject in a book of 1889, *Stellar Evolution and Its relation to Geological Time*. The idea that the stars were crucibles of "proto-elements" unknown on Earth was widespread and not confined to astronomers, physicists and chemists. Impressed by the discovery of the electron and radioactivity, in 1908 George F. Becker, a geologist of the US

---

[26] On Einstein's hypothesis and some other suggestions of charge inequality in a cosmological context, see Kragh (1997). Around 1960 Raymond Lyttleton, Hermann Bondi and Fred Hoyle developed an "electrical universe" model on this basis, arguing that it amounted to strong support of a steady-state universe. At the time neither of them drew geophysical consequences from cosmological theory, but they would do so about a decade later (see Sect. 4.1).

[27] See the historical bibliography in http://www.aip.org/history/climate/bibdate.htm

Geological Survey in San Francisco, published a paper on "Relations of Radioactivity to Cosmogony and Geology." Becker distinguished between elements found on the Earth and those found in meteorites, the Sun, the stars and the nebulae. He concluded that by charting in this way the distribution of the elements the plausibility of the evolution hypothesis was increased and that his table pointed to "the truth of the hypothesis that elements are evolved."[28]

The first attempts to establish finite-age physical cosmologies appeared about three decades later. These attempts, and especially those proposed by the German physicist Carl Friedrich von Weizsäcker in the late 1930s, relied on the assumption that the chemical elements were formed by nuclear processes in the early universe. The abundance distribution of elements or isotopes was therefore of great importance to this kind of theory, which in the late 1940s was developed by George Gamow in particular. The main test for Gamow's nuclear-archaeological theory was provided by the cosmic abundance data published in 1938 by the Norwegian geologist and geochemist Victor Goldschmidt.[29]

The link between geochemistry and cosmology appears clearly in a textbook written by two Finnish geologists, Kaleva Rankama and Ture Sahama, which included a substantial chapter on nuclear physics and cosmology in relation to the formation and distribution of the elements. While the explosion theory of Gamow and his collaborators did not attract much attention outside a small group of nuclear physicists, it was cautiously accepted by the two geologists according to whom "the nuclides were not formed as a result of a frozen equilibrium state but rather in a continuous, unfinished building-up process from a highly compressed overheated neutron gas, which decayed into protons and electrons."[30] What Rankama and Sahama described was an early version of the big bang.

1.3 Halm's Expanding Earth

German-born Jakob K. E. Halm, a respected British-South African astronomer, originally specialized in solar spectroscopy. After having moved from Edinburgh to the Cape Observatory in South Africa, in 1911 he published an important paper in which he suggested, as the first astronomer, a relationship between the mass of stars and their luminosity and evolutionary state. This was the first version of the mass-luminosity relation which was only fully clarified by Eddington thirteen years later. Still in 1917 most astronomers thought that interstellar absorption was negligible, but this year Halm concluded from studies of the distribution of stars

---

[28] Becker (1908), p. 145. See also Kragh (2000).

[29] On Goldschmidt's research programme and the transition from geochemistry to cosmochemistry, see Kragh (2001).

[30] Rankama and Sahama (1950), p. 70.

that light was absorbed along the galactic equator at a maximum amount corresponding to 2.1 magnitudes per kiloparsec.[31]

According to Halm, the evolutionary history of the Earth was a problem that invited astrophysical and not merely geophysical consideration. In a remarkable paper of 1935 based on his presidential address to the South African Astronomical Association, he used astrophysical reasoning to argue that the Earth was expanding, contrary to the accepted view of a static or slightly contracting Earth. His approach to the Earth was distinctly astrophysical and entirely different from the traditional geological approach, which he criticized for being too limited and based on the axiom of a slowly contracting Earth. Halm insisted that the evolution of the Earth could only be understood on the basis of astrophysical theory and that such a perspective inevitably led to a very different picture, namely that the Earth had expanded through its entire history.

From thermodynamic considerations of stellar and planetary atmospheres Halm obtained an equation which gave an invariant relation between the mean absolute temperature T_s of a planet's (or star's) surface and its mean density ρ measured in the unit g cm^{-3}:

$$\frac{T_s}{\rho^{8/21}} = \text{constant} \, (C)$$

We have found, he wrote, that

> ... for every star there comes a moment when its life as an active gaseous body comes to an end. For reasons, the meaning of which we have not yet grasped, it is turned abruptly into a *rigid* body, the *rigor mortis* of star life has set in. ... Once this rigor mortis has set in, the further fate of these star corpses is clearly defined by the condition in [the above] equation. The star *expands*, and the ratio between cooling and expansion is strictly regulated in accordance with [this] equation.[32]

Using available data for stellar and planetary surface temperatures, he noted that white dwarfs and planets were characterized by approximately the same value, namely $C = 145$. The "fundamental equation" or "equation of evolution," as he also called it, could thus be written

$$T_s = 145 \times \rho^{0.38}.$$

Halm emphasized that the equation was valid for all celestial bodies ranging from white dwarfs to the coolest planets. From the temperature-density law and certain speculative assumptions concerning the size of atoms at very high pressure he derived that "at the beginning of geological time" the radius of the Earth was 5430 km and its mean surface temperature about 63 °C. The radius R of the primitive Earth would thus be less than the present one by 941 km or "about

---

[31] On Halm's life and work, see Glass (2014).

[32] Halm (1935a), p. 14.

100 times the height of Mount Everest." As to the average rate of expansion he estimated it to be $dR/dt \cong 1.6$ mm years^{-1} or "about the thickness of a penny-piece."

Since Halm assumed the Earth's mass to remain constant, in the geological past the density and surface gravity of the Earth would have been considerably higher than the present values. He estimated the original density to 9.13 g cm^{-3} or 3.46 g cm^{-3} larger than today. As to the surface temperature at two different epochs at which the radius of the Earth was R_1 and R_2, respectively, he calculated

$$\frac{T_1}{T_2} = \left(\frac{R_2}{R_1}\right)^{1.14}.$$

For the past climate it meant that an increase in radius of 1 % corresponded to a 3.7 °C lower temperature. Contrary to later expansionists Halm argued that the primitive Earth was entirely covered by water, the continents only arising along with the expansion of the Earth.

Halm's theory of the Earth as a slowly cooling and expanding rigid body offered a new perspective on "the remarkable and fascinating suggestion regarding the formation of the continents made recently by the German geologist Wegener." At the time Alfred Wegener's theory of continental drift was not highly regarded and rejected by a majority of geologists and palaeontologists. The major supporter of drift in the 1930s was Halm's compatriot, the South African geologist Alexander du Toit. Halm's theory evidently differed from Wegener's, which assumed horizontal plate displacements on an Earth of constant size. While Halm admitted that there was no adequate physical force to move the continents in accordance with the drift theory, he nonetheless thought that Wegener's basic idea "is so strongly supported by their [the continents'] present configurations that it cannot be lightly rejected."

In agreement with Wegener, Halm believed that the continents derived from a common supercontinent (Pangaea), only had this original continent split as a result of the expansion. He did not think of the expanding Earth as an alternative to continental drift, but rather as an improved version of it. "The single conception of the Earth as an *expanding* body," he wrote, "has based Wegener's fascinating theory on a sound physical principle and has opened new vistas of approach towards the solution of the many problems which the history of our planet lays before us."

Halm's theory attracted very little attention, but some later proponents of expansionism considered it a precursor of what in the 1960s emerged as the modern expansion theory of the Earth and which will be dealt with in Chaps. 3 and 4.[33] Although by 1935 the expansion of the universe was well known among astronomers, Halm did not suggest a connection between planetary expansion and cosmic expansion such as a few later expansionists would do. Indeed, he did not refer to the latter phenomenon at all, and this for the simple reason that he did not believe in the theory of the expanding universe. As he made clear in a companion article, he much

---

[33] Carey (1988), p. 140.

preferred a classical explanation of the redshifts based on the static universe over the "maze of abstruse speculations" that characterized the relativistic explanation.[34] Halm consequently devised an explanation of the linear redshift-distance relation which built on the assumption that an undisturbed wave of light may undergo adiabatic expansion without violating energy conservation. Contrary to other "tired-light" hypotheses of the period, Halm's explanation did not appeal to inter-action between light and matter, or between light and gravitational fields. Quantum theory and Planck's constant did not appear in his scheme, which was based solely on a classical analysis of wave motion.

---

[34] See Halm (1935b), which is reprinted in Kragh (2015d). The first tired-light hypothesis accounting for Hubble's redshift-distance relation was proposed by Fritz Zwicky as early as August 1929, before the idea of an expanding universe was generally known.

Chapter 2
Varying Gravity

The unorthodox idea that the gravitational constant G varies slowly in time arose in the late 1930s in the context of a cosmological theory proposed by the English physicist Paul Dirac. The idea was received coolly, not only because it led to a much too small age of the universe but also because it contradicted the general theory of relativity and, on the top of that, was thought to be untestable. However, Dirac's idea was taken up and further developed by Pascual Jordan in Germany and after World War II it slowly began to attract attention among physicists and astronomers. In 1948 the hypothesis of varying gravity made its first connection to the earth sciences in the form of an attempt, made by Edward Teller, to test the hypothesis by means of a paleoclimatic argument. The test was inconclusive and was initially ignored by the earth scientists.

Although varying gravity has no place in Einstein's general theory of relativity it is part of the "scalar–tensor" theories developed by Jordan, Carl Brans, Robert Dicke and others. This chapter focuses on the contributions of Dirac, Jordan and Dicke and their arguments for a gravitational constant decreasing over cosmic time. In so far that these arguments were of a geophysical nature they will be dealt with in more detail in Chap. 3.

2.1 Big G: The Gravitational Constant

The idea that the constant of gravitation G is not a true constant but varies in time dates from the 1930s. It will be useful to look briefly at the earlier history of the constant and the law of nature with which it is so firmly associated. Newton's famous law of universal gravitation dates back to Book III of the *Principia* from 1687. According to the standard formulation of this law—which was not Newton's formulation—two masses m and M separated by the distance r attract one another by a force F given by

© Springer International Publishing Switzerland 2016
H. Kragh, *Varying Gravity*, Science Networks. Historical Studies 54,
DOI 10.1007/978-3-319-24379-5_2

$$F = G\frac{mM}{r^2}.$$

Here G is a gravitational constant with the same value everywhere in the universe. Although Newton proposed the law on the basis of our solar system alone, he boldly postulated its universal validity.

Newton's law of gravitation is the first fundamental law of nature ever in the history of science and it quickly came to be recognized as the foundation of celestial mechanics and other branches of theoretical astronomy. It also played a crucial role in early geodesy, geophysics, and gravity surveying. From a modern point of view the important part of Newton's law is "big G," one of the first of the fundamental constants of nature which are still today assigned this divine status. The first such constant was the velocity of light $c \cong 3 \times 10^8$ m s^{-1} discovered in 1676 but only recognized as a universal constant much later. In modern parlance big G is sometimes referred to as the "gravitational coupling coefficient" and stated in terms of a very small dimensionless quantity, namely

$$\alpha_{\mathrm{G}} = G\frac{2\pi m^2}{hc} \cong 1.8 \times 10^{-45},$$

where m and h denote the electron's mass and Planck's constant, respectively. However, one will look in vain for G or the term "gravitational constant" in *Principia* or in other works from the early period. Newton and his successors in the age of the Enlightenment saw the law of gravitation as a relation between proportions and not between absolute quantities. In astronomical contexts the absolute value of G was irrelevant, since only Gm or GM was important.[1] The goal of experimenters in the eighteenth and nineteenth centuries was not to determine G but primarily to determine the mean density of the Earth, a quantity for which the absolute value of G played no role.

Textbooks tell us that the English natural philosopher Henry Cavendish in a celebrated experiment published in 1798 determined the value of G (Fig. 2.1). He did not and had no intention of doing so. That Cavendish tried to measure G is nothing but a myth, a textbook anachronism. He *could* have calculated G from the measured density ρ of the Earth, its radius R and the surface acceleration g, which quantities (assuming the Earth to be spherical) are related as

$$G = \frac{3}{4\pi}\frac{g}{\rho R}.$$

But Cavendish did not refer to G and neither did other physicists at the time. The

---

[1] On pre-relativistic conceptions of the gravitational law and its associated constant, see Will (1987), Ducheyne (2011), and Uzan and Lehoucq (2005), pp. 253–261. A detailed account of experiments on gravitation until the 1890s is given in Mackenzie (1900). References to the literature can be found in these sources.

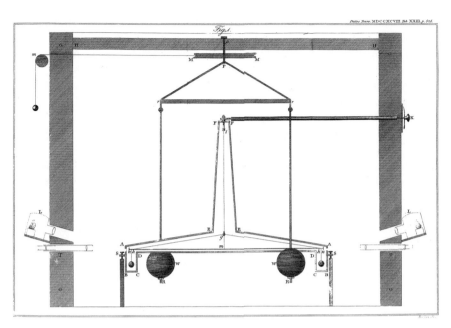

Fig. 2.1 Cavendish's torsion-balance apparatus designed to measure the mass of the Earth or retrospectively the gravitational constant. *Source*: *Philosophical Transactions of the Royal Society* (1798), p. 526

title of his paper—his last one—was "Experiments to Determine the Density of the Earth." Using an improved version of an apparatus originally designed by John Michell, an English clergyman and astronomer, Cavendish determined "to great exactness" the mean density of the Earth to be 5.48 times the density of water.[2] There was nothing in his analysis to require the gravitational constant, a quantity which was of no relevance to his determination. Had Cavendish calculated *G* from the experimental data obtained with his torsion balance he would have obtained 6.71×10^{-11} m^3 kg^{-1} s^{-2}, a remarkably good result less than 1 % from its present value.

Only in 1873 did two French physicists, Marie-Alfred Cornu and Jean-Baptiste Baille, point out the importance of determining *G* in absolute terms. The title of their paper referred specifically to "the constant of attraction" for which they used the symbol *f* which at the time was common also among German physicists and astronomers. For the mean density of the Earth the two Frenchmen arrived at the value (5.530 ± 0.030) g cm^{-3}.

It took another two decades until *f* or *G* became an established object of scientific inquiry and soon regarded as more important than the Earth's density. The British physicist John H. Poynting at the Mason Science College, Birmingham, was among the first to use the symbol *G*, argue for its fundamental nature, and actually

---

[2] For details on Cavendish and his experiment, see Jungnickel and McCormmach (1999), pp. 440–460.

measuring its value. In a paper of 1892 he reported $G = 6.6984 \times 10^{-11}\,\mathrm{m^3\,kg^{-1}\,s^{-2}}$ corresponding to a mean density of the Earth of $\rho = 5.4934$ g cm^{-3}. Three years later his compatriot Charles V. Boys, a physicist at the Royal College of Science in London,   concluded   from   a   series   of   careful   experiments   that $G = 6.6579 \times 10^{-11}\,\mathrm{m^3\,kg^{-1}\,s^{-2}}$ and $\rho = 5.5268$ g cm^{-3}. The presently accepted value of Newton's constant is

$$G = (6.67384 \pm 0.00080) \times 10^{-11}\,\mathrm{m^3\,kg^{-1}\,s^{-2}}.$$

The quantity still remains the most difficult of the fundamental constants to measure and the one known least accurately. This is not only because gravity is such a weak force but also because the experimental apparatus cannot be shielded from the gravitational influences of other bodies. While G is considered an extremely important quantity, g (9.8067 m s^{-2}) and the mean density of the Earth ρ (5.515 g cm^{-3}) is today of limited interest only.

The change in attitude in the late nineteenth century was spelled out by Boys in a lecture of 1894 to the Royal Institution, where he described g to be merely "the delight of the engineer and the practical man." By contrast, G "represents that mighty principle under the influence of which every star, planet, and satellite in the universe pursues its allotted course." Boys continued:

> It is a mysterious power, which no man can explain; of its propagation through space, all men are ignorant. It is in no way dependent on the accidental size or shape of the earth; if the solar system ceased to exist it would remain unchanged. I cannot contemplate this mystery, at which we ignorantly wonder, without thinking on the altar on Mars' hill. When will a St. Paul arise to declare it to us? Or is gravitation, like life, a mystery that can never be solved?[3]

Some two decades later, a St. Paul did arise. Referring to Cavendish's experiment, Boys declared: "Owing to the universal character of the constant G, it seems to me to be descending from the sublime to the ridiculous to describe the object of this experiment as finding the mass of the earth or the mean density of the earth."

On the theoretical side a prescient paper by the Irish physicist George Johnstone Stoney deserves mention.[4] In 1881 Stoney highlighted the significance of G by raising it to a natural constant on par with other constants such as the elementary charge e and the speed of light c. As to e, he derived the value from a corpuscular interpretation of electrolytic experiments and proposed ten years later to call the elementary charges "electrons." By combining the three constants he proposed new fundamental units for time, length and mass. For example, the Stoney unit for length was the inconceivably small number

---

[3] Boys (1894), p. 330.

[4] Stoney (1881). See also Barrow (2002), pp. 18–23.

$$l_S = \frac{e}{c^2} \sqrt{G} \cong 10^{-37}\,\text{m}.$$

The Stoney units were essentially the same as the better known Planck units first introduced by Max Planck in 1899, at a time when he was not yet in possession of the famous *h* constant named after him. The Planck length is 1.6×10^{-35} m. At about the same time there were several attempts to explain gravitation in terms of electron theory or electromagnetism generally. However, none of the attempts came even close to success and none of them questioned the constancy of *G*. On the other hand, they resulted in the first values of the dimensionless ratio between the gravitational and the electromagnetic interactions which a few decades later would inspire Dirac to suggest a time-varying *G*. In 1882, at a time when the electron had not yet been either named or discovered, the German astrophysicist Carl Friedrich Zöllner at Leipzig University derived the relation

$$\frac{e^2}{Gm^2} \cong 3 \times 10^{40}.$$

The quantity *m* refers to the mass of a hypothetical particle with the elementary charge $\pm e$. With the emergence of electron theory at the end of the century several physicists investigated possible connections between electricity and gravitation. Given that both forces vary inversely with the square of the distance it was tempting to suppose some kind of connection. One of the physicists was the American Bergen Davis at Columbia University, who in 1904 found the approximate value 8×10^{41} for Zöllner's ratio. He based his derivation of the connection between *G* and "the electrical constants of the ether" on the popular electromagnetic world view. The early contributions of Zöllner and Davis are forgotten today, when the dimensionless number of the order 10^{40} is often referred to as either Weyl's or Eddington's number, referring to the German mathematician Hermann Weyl and the British astronomer Arthur Eddington, respectively.[5]

The aim of most laboratory experiments on gravitation in the late nineteenth century and the early twentieth century was to test if Newton's law was valid at relatively small distances and if it was independent of, for example, the composition of the masses or of the intervening medium. In an address of 1900, Poynting stated that "the attempts to show that, under certain conditions, it [*G*] may not be constant, have resulted so far in failure all along the line."[6] With these words Poynting did not refer to the possibility that *G* might vary in time, a conjecture which to my knowledge was absent from physics until the 1930s. Given that until that time there was no generally accepted time frame of the universe this is perhaps understandable. What could *G* possible vary with? It was only with the recognition

---

[5] Davis (1904). For Zöllner's number, see Kragh (2012). Weyl first discussed the number 10^{40} in 1919 and Eddington in 1923.

[6] Poynting (1920), p. 643. See also Ducheyne (2011), p. 2011.

of a cosmological arrow of time in the form of the expansion of the universe that the question of a time-varying gravitational constant arose.

2.2 Dirac and the Magic of Large Numbers

In the late 1930s cosmological models based on the expanding solutions of Einstein's field equations were gaining ground.[7] The canonical tensor equations dating from 1917 and including the cosmological constant Λ are

$$R_{\mu\nu} - \frac{1}{2}g_{\mu\nu}R - \Lambda g_{\mu\nu} = -\kappa T_{\mu\nu}.$$

The equations express how the geometry of space-time relates to the content of matter and energy as given by the energy-momentum tensor $T_{\mu\nu}$ ($\mu,\nu = 1, 2, 3, 4$). The quantity $R_{\mu\nu}$ is known as the Ricci tensor and R is a curvature invariant derived from $R_{\mu\nu}$; the components of the fundamental tensor $g_{\mu\nu}$ are functions of the chosen coordinates. Finally, the symbol κ denotes the Einstein gravitational constant which is related to the Newtonian constant of gravitation G as

$$\kappa = \frac{8\pi}{c^2}G.$$

Although a minority of astronomers still defended a static picture of the universe, the majority agreed that the universe is in a state of expansion and that the recession rate of the galaxies corresponded to a value of the Hubble constant given by $H_0 \cong 500$ km s^{-1} Mpc^{-1} (1 Mpc $= 10^6$ parsec $\cong 3.1 \times 10^{22}$ m).

Lemaître's suggestion of a finite-age universe originating from a highly compact "primeval atom" dates from 1931, but this kind of model was considered unattractive by most physicists and astronomers. To the extent there was a favoured model of the universe, it was rather the Lemaître–Eddington model that assumed a gradual expansion from a pre-existing static Einstein state. Moreover, not all agreed that general relativity was the natural or only framework of cosmology. Edward Arthur Milne's alternative "kinematic theory" rested on very different conceptions of space and time, including that space was just an abstract reference system and therefore could have no curvature. For a decade or so Milne's system attracted much interest among British astronomers and physicists in particular (see further in Sect. 2.4). Some of Milne's ideas would continue to influence cosmological thinking decades after his death in 1950, but today they are largely forgotten.

The British theoretical physicist Paul Adrien Maurice Dirac was a Nobel laureate of 1933 and famous for his fundamental contributions to quantum mechanics, a

---

[7] The history of cosmology in the period is the subject of several books, including North (1965), Kragh (1996), and Nussbaumer and Bieri (2009).

Fig. 2.2 Paul A. M. Dirac.
Credit: Niels Bohr Archive,
Copenhagen

theory he co-founded with Werner Heisenberg and others in 1925 (Fig. 2.2). Three years later he established a quantum wave equation for the electron in agreement with the special theory of relativity and on the basis of this equation he predicted the existence of the "antielectron" soon to be known as the positron. Although Dirac's entire work up to the mid-1930s had been in quantum theory he also had an interest in Einstein's general theory of relativity and its cosmological consequences. Indeed, he was one of the first physicists to adopt Lemaître's daring picture of a universe with a violent beginning in time. In the spring of 1933 Dirac listened to a talk Lemaître gave on his "primeval atom" theory in Cambridge and after the talk he discussed the problems of cosmology with the Belgian physicist and priest.

Inspired by Lemaître, Milne and Eddington, in a brief note in *Nature* of February 1937 Dirac discussed the significance of two large dimensionless combinations of constants of nature. In the CGS units used at the time the combinations were

$$\frac{T_0}{e^2/mc^3} \cong 2 \times 10^{39} \text{ and } \frac{e^2}{GmM} \cong 7 \times 10^{38}.$$

The symbols e, c and G denote the elementary charge, the velocity of light in vacuum and Newton's gravitational constant, respectively. M is the proton's mass and m the electron's mass, and T_0 is the age of the universe since the "beginning about 2×10^9 years ago, when all the spiral nebulae were shot out from a small region of space, or perhaps from a point."[8] Dirac thought that the two very large numbers of the order 10^{39} or 10^{40} must be related in a simple way, meaning that the first expression must be roughly equal to the second. Further assuming that the constants m, M, e and c are truly constant, it follows that the gravitational constant decreases inversely with cosmic time:

---

[8] Dirac (1937).

$$G \sim \frac{1}{t} \quad \text{or} \quad \frac{1}{G}\frac{dG}{dt} \sim -\frac{1}{t}.$$

In his note of 1937 Dirac also referred to the quantity

$$N = \frac{\rho}{M}\left(\frac{c}{H}\right)^3 \cong 10^{78},$$

where c/H is the Hubble distance and ρ the average density of matter in the universe, which he took to be 5×10^{-31} g cm^{-3}. As Eddington had first observed, the quantity is a measure of the number of protons (or nucleons) in the visible universe. According to Dirac, it was no coincidence that this number N is close to the square of the age of the universe measured in atomic time units. It implied, he suggested, that protons would be created as the universe grew older, following

$$N \sim t^2$$

Dirac proposed as a general principle that when two very large numbers of the order 10^{39} and 10^{78}—or generally $(10^{39})^n$, where n is a natural number—occur in nature, they must be connected by a simple mathematical relation. This became known as the *Large Numbers Hypothesis* (LNH), a name he only introduced in the 1970s and which since then has stuck.[9]

The essence of Dirac's hypothesis was that the laws of physics are evolutionary, such as he stressed in a lecture he gave in Edinburgh in February 1939:

> At the beginning of time the laws of Nature were probably very different from what they are now. Thus we should consider the laws of Nature as continually changing with the epoch, instead of as holding uniformly throughout space-time. ... As we already have the laws of Nature depending on the epoch, we should expect them also to depend on position in space, in order to preserve the beautiful idea of the theory of relativity that there is a fundamental similarity between space and time.[10]

In a follow-up paper to his note in *Nature* of 1937 Dirac developed his numerological arguments into a quantitative cosmological model with testable consequences. However, he now decided to drop the idea of accelerating creation of matter with the sole argument that "there is no experimental justification for this

---

[9] Dirac (1938) referred to the "Fundamental Principle." He first spoke of the Large Numbers Hypothesis in a talk given in 1972. See Dirac (1973a), p. 46. Pascual Jordan always referred to "Dirac's principle." For more details on the history of the principle and Dirac's cosmology, see Kragh (1990), pp. 223–246 and also Barrow and Tipler (1986). Many of Dirac's arguments concerning the nature and use of the LNH were questionable and based on somewhat arbitrary assumptions. See, for example, Klee (2002) for an interesting critical review. However, from the point of view of the present book this is less relevant.

[10] Dirac (1939), p. 128.

assumption."[11] From the LNH he obtained for the scale factor R that it increased in time according to

$$R(t) \sim t^{1/3}.$$

This implied a deceleration parameter of

$$q_0 \equiv - \left(\frac{RR''}{R'^2} \right)_0 = 2.$$

The symbols R' and R'' denotes dR/dt and d^2R/dt^2, respectively. While at the time a deceleration as high as $q_0 = 2$ was not ruled out by observation, the predicted age of the universe was highly problematic. From Dirac's theory it follows that the present age relates to the Hubble constant H_0 according to

$$t_0 = \frac{T_H}{3} = \frac{1}{3} H_0^{-1}.$$

As Dirac noted with an understatement, this gave the "rather small" age of 7×10^8 years or less than one third of the age of the Earth as determined by radioactive methods. Dirac's value of $t_0 = 7 \times 10^8$ years or $T_H = 2.1 \times 10^9$ years is a little puzzling since it corresponds to a Hubble constant $H_0 = 465$ km s^{-1} Mpc^{-1}. In 1938 the accepted value, as determined by Edwin Hubble, was $H_0 = 540$ km s^{-1} Mpc^{-1} or $T_H = 1.8 \times 10^9$ years. No astronomical measurements indicated a value less than 500 km s^{-1} Mpc^{-1}. Dirac gave no source for his value. Instead of considering the disagreement between the two time-scales a mortal blow against his theory he vaguely suggested that the problem might be solved if the rate of radioactive decay—on which geochronology relied—decreased in cosmic time.

Dirac further argued that the only geometry compatible with the LNH was flat or Euclidean space and also that the LNH ruled out a non-zero cosmological constant. He thus ended up with a model somewhat similar to the model that Einstein and Willem de Sitter proposed in 1932 and in which the universe expands as

$$R(t) \sim t^{2/3}.$$

But whereas the Einstein–de Sitter model was a special case of the cosmological field equations, Dirac's was not. He realized that the assumption of $G(t)$ was incompatible with general relativity.

For the sake of completeness it should be mentioned that the American-Israeli physicist Samuel Sambursky independently of Dirac considered the possibility of a decreasing G.[12] However, the main purpose of Sambursky's paper was to offer a

---

[11] Dirac (1938), p. 204.
[12] Sambursky (1937).

static alternative to the expanding universe based on the assumption that Planck's constant *h* decreases exponentially with time. He suggested that a larger *G* in the past implied that old stars had originally had a lower mass than ordinarily assumed.

Dirac's cosmological model of 1938 resulted in two testable consequences. First, the age of the universe was only about 700 million years; second, the gravitational constant decreased according to

$$\frac{1}{G}\frac{dG}{dt} = -3H_0 = -\frac{3}{T_H}.$$

With the value of H_0 accepted at the time it meant a relative change of about 10^{-10} per year, which was thought to be detectable only in principle. In his paper of 1938 Dirac did not mention this second consequence explicitly, but it follows directly from his theory. In the subsequent discussions of Dirac's theory the $G(t)$ hypothesis was in focus, whereas his expanding cosmological model received almost no attention. The possibility of testing $G(t)$ indirectly by means of its consequences with respect to the Earth in the past was not mentioned by Dirac or others before Edward Teller's paper of 1948 to be considered in Sect. 2.5.

With only five citations in the period 1938–1947 (according to Web of Science) Dirac's cosmological paper of 1938 was not an immediate success (Fig. 2.3). He could not possibly have foreseen the rich and diverse literature his theory would eventually give rise to. Until 1978 Dirac's paper received 165 citations in scientific journals, a number which can be taken as a rough indication of the popularity of the $G(t)$ hypothesis in the period. Interestingly, a substantial part of the citations appeared in the context of the geological sciences (Table 2.1).

In a letter of 1967 Gamow recalled Dirac's suggestion of a varying *G*: "The first criticism of this idea was made by [Niels] Bohr. I still remember him coming to my room (I was visiting Copenhagen at that time) with the fresh issue of *Nature* in his

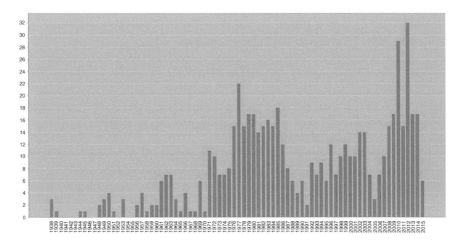

Fig. 2.3 Citations to Dirac (1938). Credit: Web of Science

Table 2.1 Citations to Dirac (1938) in the period 1938–1978

Main subject of journals and papers	Number of citations
Physics	69
Astrophysics and cosmology	63
Earth and planetary sciences	33

Source: Web of Science

hands, saying: 'Look what happens to people when they get married.'"[13] Dirac's unorthodox theory was criticized on methodological grounds, but attracted little scientific interest among physicists and even less among astronomers. Geophysicists and other earth scientists either ignored it or, more likely, were unaware of it.[14] According to the astrophysicist and philosopher Herbert Dingle, the $G(t)$ theory was a regrettable example of a "pseudo-science of invertebrate cosmythology."[15] Eddington was less vehement in his rejection, but not much kinder. Although part of Dirac's theory was inspired by Eddington's ideas, he dismissed the theory as "unnecessarily complicated and fantastic."[16] Varying constants of nature were anathema to Eddington.

On the other hand, a few astrophysicists responded with interest to Dirac's hypothesis which they speculated might be useful in understanding the interior of stars. Inspired by Dirac's note in *Nature*, Subrahmanyan Chandrasekhar suggested that the number of particles in a star might vary as $t^{3/2}$.[17] Daulat Singh Kothari, another Indian astrophysicist, applied Dirac's reasoning to the theory of white dwarf stars,[18] and in California the astronomer Fritz Zwicky did the same in the case of the even denser neutron stars. As Zwicky pointed out, "[Dirac's] hypothesis must naturally be regarded as mere speculation unless it allows us to draw further conclusions which are in agreement with observational facts."[19] While Zwicky thought that such conclusions might be relevant to astrophysics neither he nor others at the time thought of conclusions related to the past of the Earth. That only came with Teller's (1948) paper on the Earth's paleoclimate and Jordan's book of 1952, where the expanding Earth and the varying-G hypothesis were first connected. It was thus quite wrong, when a later geologist wrote that "Earth

---

[13] Gamow to the American geophysicist Philip H. Abelson, 1 September 1967, reproduced in Gamow (1967b), p. 767. On 2 January 1937 Dirac married Margit Wigner Balasz, the sister of the Hungarian-American physicist Eugene Wigner.

[14] The leading geophysicist Harold Jeffreys was aware of Dirac's hypothesis as early as June 1937, when he commented on the hypothetical-deductive methodology of Dirac, Milne and Eddington (see *Nature* **141**: 1004–1006). However, neither at this nor at later occasions did he mention the $G(t)$ hypothesis and its possible geological consequences.

[15] Dingle (1937). See Kragh (1996), pp. 69–71 for Dingle's attack on Dirac's theory and other fundamental theories which he accused of being rationalistic fantasies with no foundation in either experiment or observation.

[16] Eddington (1939), p. 234.

[17] Chandrasekhar (1937).

[18] Kothari (1938).

[19] Zwicky (1939), p. 733.

scientists have been conscious of the possibility of Earth expansion ever since Dirac
... suggested that the gravitational constant may be slowly decreasing."[20]

Among the few physicists who valued Dirac's hypothesis was the young
Frenchman Jacques Solomon, who discussed Dirac's article of 1938 shortly after
it had appeared.[21] In an early investigation of quantum gravity Solomon mentioned
that together with Planck's constant h and the proton's mass M the gravitational
constant defined a length unit given by

$$\frac{h^2}{GM^3} \cong 10^{27} \text{ cm.}$$

This length, he pointed out, was of roughly the same order of magnitude as the
radius of the static Einstein universe as given by

$$R_\mathrm{E} = \frac{1}{2} \frac{c}{\sqrt{\pi G}} \frac{1}{\sqrt{\rho}}.$$

Noting that the Fermi constant G_F in the theory of beta radioactivity was about
10^{-13} (if expressed in the unit erg), Solomon further suggested that, according to
the LNH, G_F might decrease with cosmic time according to

$$G_\mathrm{F} \sim \frac{1}{t^{1/3}}.$$

Since the decay constant λ_β is proportional to the square of G_F the suggestion
implied that beta-active substances would decay at a different rate in the past than
presently. Solomon argued that in this way the age of the Earth might be reconciled
with the age of the universe.

2.3   Jordan's Cosmological System

The eminent German physicist Ernst Pascual Jordan had in 1925 been a co-founder
of the new quantum mechanics and a few years later, together with Dirac, he
established the foundation of quantum field theory (Fig. 2.4). Between 1929 and
1944 Jordan worked as professor at the University of Rostock. During this period he
increasingly turned toward less mainstream areas of science, including cosmology
and what he called "quantum biology," an ambitious attempt to understand life on
the basis of quantum physics. Adding to his diminished scientific reputation and

---

[20] Smith (1978).

[21] Solomon (1938). Solomon, a Jew and a communist, was active in the French resistance
movement during the early years of World War II. He was executed by the Germans in 1942.

Fig. 2.4 Pascual Jordan
(*right*) in conversation with
Werner Heisenberg (*left*)
and Max Delbrück in
Copenhagen 1936. Credit:
Niels Bohr Archive,
Copenhagen

estrangement from mainstream physics was his strong support for Hitler's Third
Reich. Jordan joined the Nazi party in 1933 and after the German defeat in World
War II he had to go through a de-nazification process. Only in 1953 did he again
become a full professor of physics, this time in Hamburg. For quite some time his
reputation was tainted by his Nazi past. For example, as late as about 1960 the
American physicist Joshua Goldberg had misgivings about supporting Jordan with
funds from the U.S. Air Force because the German physicist "had been close to the
Nazi government."[22]

"I am the only one who has been ready to take Dirac's world model seriously ...
and to consider its more precise formulation," Jordan rightly commented in a book
of 1952. He described Dirac's LNH and the resulting $G(t)$ hypothesis as "one of the
great insights of our time."[23] Until captivated by Dirac's LNH Jordan had worked
almost exclusively in quantum theory and paid no attention to the problems of
gravity and general relativity. The meeting with Dirac's hypothesis changed his
research track dramatically, such as he recalled in a report from 1961:

> My interest in general relativity arose when I became acquainted with Dirac's cosmological
> speculations, which lead him in 1937 to his hypothesis about cosmologically diminishing
> gravitation. I have been then nearly the only one among physicists who was really
> fascinated by Dirac's idea and convinced by his arguments, and tried to think more about
> them. ... I felt myself necessitated to study more thoroughly general relativity, a branch of
> modern physics to which before these events I had devoted only cursory attention, having
> been busied with the fascinating problems of quantum mechanics or quantum field theory—
> which I had to lay aside now in order to study gravitation.[24]

---

[22] Goldberg (1992), p. 95. Goldberg was a leading physicist in the Aeronautical Research Laboratories (ARL) established in 1955 by the U.S. Air Force. For Jordan and ARL, see Sect. 2.6.
Jordan was not invited to participate in the 1958 Solvay Congress on gravitation and cosmology.
One may speculate that one of the reasons was that his past as a Nazi was not easily forgotten in
formerly occupied Belgium.

[23] Jordan (1952), p. 137.

[24] Jordan (1961b), p. 2.

Jordan shared with Dirac that he was an early convert to Lemaître's picture of an exploding universe, which he endorsed even before he got involved with Dirac's LNH and its associated cosmological ideas. In a semi-popular book of 1936 Jordan wrote as follows:

> Ten billion years ago—Lemaître especially deserves credit because of the closer execution of this representation—the initially small universe arose from an original explosion. Not only atoms, stars and Milky Way systems but also space and time were born at that time. Since then the universe has been growing, growing with the furious velocity which we detect in the flight of the spiral nebulae.[25]

In a series of papers starting the following year Jordan developed his own cosmological system that incorporated some of the numerological ideas of Eddington and Dirac, including the assumption of a decreasing gravitational constant.[26] However, contrary to Dirac he maintained the conclusion of a spontaneous creation of matter following $N \sim t^2$. In a booklet of 1947 titled *Die Herkunft der Sterne* (The Origin of the Stars), Jordan gave a comprehensive survey of his cosmological and astrophysical ideas based on what he called "Dirac's principle." A popular summary account appeared the same year.[27]

Rather than hydrogen atoms being formed individually and continually throughout space, Jordan proposed the more radical idea that entire stars and galaxies were formed ex nihilo along with the expansion of space.[28] While the mass of the universe increased as t^2, he concluded that the mass of an individual star would increase as $t^{3/2}$. This was not to be understood in the sense that a star continuously became heavier, but in the sense that the later in the history of the universe a star was formed the heavier it would be.

According to Jordan, a star would start its life as a spontaneously created droplet of matter with density equal to that of an atomic nucleus, $\rho \cong 2 \times 10^{14}$ g cm^{-3}. Such a nuclear droplet he identified with a supernova, which initially would have a radius of only 1 mm. (There is some similarity between Jordan's ideas and the later concept of the hypothetical "white holes" as time-reversed black holes.) However, as pointed out by Hermann Bondi, the rate of supernova formation following from the scenario was much too high.[29] Although Jordan's picture of the stellar universe was clearly speculative, he maintained that it agreed with "the epistemological

---

[25] Jordan (1936), p. 152. The book appeared in an English translation as *Physics of the 20th Century* (New York: Philosophical Library, 1944). On Jordan's cosmological theories and references to his work in this area, see Kragh (2004), pp. 175–185.

[26] Jordan (1937, 1939, 1944).

[27] Jordan (1947a, b).

[28] As expressed by Singh (1970), p. 233, in Jordan's theory stars came "literally out of the blue like Athena leaping forth from Zeus's brain mature and in complete armour." On the request of his former professor Max Born, Jordan wrote a summary article in English of his work in cosmology and astrophysics. See Jordan (1949) and also the critical review of Jordan's theory in North (1965), pp. 205–208.

[29] Bondi (1952), p. 164.

requirements of positivism."[30] Perhaps it did, but if so it was not enough to convince the majority of physicists and astronomers of its soundness. Very few shared the view of Paul Couderc, an astronomer at the Paris Observatory, according to whom Jordan's cosmology was of "undeniable elegance."[31]

The world model that Jordan arrived at was indebted to Lemaître as well as to Dirac, but he developed it in his own way. The correspondence between the research interests of Jordan and Dirac is remarkable. Not only did the two physicists develop much of quantum electrodynamics and quantum field theory in parallel, Jordan (collaborating with Eugene Wigner) also came close to the relativistic wave equation of the electron that Dirac published in early 1928. They both accepted Lemaître's model of the universe at an early time. When Dirac proposed the existence of magnetic monopoles in 1931, Jordan was the only physicist to take up the theory and develop it further. The same kind of interaction at a distance occurred in the case of Dirac's $G(t)$ cosmology.

According to Jordan there once had been an *Urexplosion* (primeval explosion)—a kind of big bang—and since then the universe had expanded linearly with time ($R = ct$), implying that the age of the universe equals the Hubble time. Like Lemaître, but contrary to Dirac, Jordan argued that cosmic space must have a small positive curvature and therefore be finite. On the other hand, he followed Dirac and most other cosmologists in putting the cosmological constant equal to zero. Jordan's *Urexplosion* was not really an explosive event of Lemaître's type, for it started in an incredibly small and empty space out of which a pair of neutrons was produced. Only with the expansion of space did matter in the form of atoms, and eventually stars and galaxies, come into existence.

The continual creation of matter might seem to rule out a description of the universe in terms of general relativity, but Jordan suggested that energy was in fact conserved during the entire cosmic process. Reviving an Eddington-inspired idea first proposed by the Austrian physicist Arthur Haas,[32] Jordan argued that the mass increase caused by the creation of new matter m was exactly compensated for by the increase in negative potential energy due to the expansion. If the mass of the universe contained within the Hubble radius $R = cT$ is denoted M the Haas–Jordan requirement can be written as

$$mc^2 - \frac{GmM}{R} = 0.$$

Thus, the universe was born with zero total energy and would remain in a zero energy state in agreement with the law of energy conservation.[33] The positive rest

---

[30] Jordan (1944), p. 190.

[31] "The construction erected by Pascual Jordan is of undeniable elegance, and at any rate suggests a simple and brilliant interpretation of the expansion." Couderc (1952), p. 225.

[32] Haas (1936). Arthur Erich Haas was at the time professor of physics at Notre Dame University in Indiana. An account of his work in speculative cosmology can be found in Kragh (2004), pp. 189–194.

[33] Jordan (1939, p. 66, 1949, p. 638).

energy of a new star mc^2 cancels the negative gravitational energy due to it.

The Haas-Jordan idea of a zero-energy universe has a curious history. It was first suggested by Haas in 1936 and then by Jordan (citing Haas) in 1939. Thirty-three years later the idea was independently reintroduced by Edward Teller, who discussed it on the basis of Dirac's cosmological theory. The following year Edward Tryon, a physicist at the City University of New York, once again "discovered" the zero-energy closed universe, this time in the context of quantum cosmology. Tryon ascribed the idea to a topological argument made by the relativist Peter Bergmann. A few years later it was "rediscovered" by S. Warren Carey (see Sect. 4.1) and in modern cosmology the idea is popular in various inflation scenarios of the early universe. Teller, Tryon, Carey as well as most other scientists seem to have been unaware of their predecessors. Based on reasoning related to Mach's principle the British physicist Dennis Sciama proposed the same idea in 1953, if in the approximate form $GM/Rc^2 \cong 1$.[34] In this form it was also discussed by several other physicists, both in relation to relativistic models and the steady-state theory. As we shall see in Sect. 2.7, the relation played an important role for Robert Dicke in his attempt to revise and extend general relativity theory.

The greater part of Jordan's cosmological and astrophysical ideas were developed during or shortly after World War II and for this reason alone they were not well known among Western scientists. However, they did attract some attention in Germany where they were taken up by several astronomers and physicists. One of them was the astronomer Paul ten Bruggencate, director of the Göttingen University Observatory, who was in contact with Jordan and had an interest in problems of a cosmological nature. In 1945 Bruggencate published a paper in the proceedings of the Göttingen Academy of Sciences in which he examined for the first time the luminosity, age and energy production of the Sun on the Dirac–Jordan assumption of $G \sim 1/t$.[35] Assuming an age of the universe of 6.5×10^9 years he calculated from Jordan's theory the plausible age of 3.6×10^9 years for the Sun. Bruggencate also calculated in his little-known paper how the Sun's luminosity would depend on its age, finding that in the past the Sun would have been considerably brighter than today. He briefly referred to the possible consequences for the history of the Earth but without recognizing that a brighter sun might cause problems with regard to the Earth's climate in the geological past.

As mentioned, in 1938 Dirac hinted that the age problem might be solved if one assumed radioactive decay to vary with the epoch. In the same year Jordan suggested that the Fermi constant of weak interactions might depend on G and therefore vary in time.[36] The idea of a possible connection between beta decay and gravitation was aired by several physicists in the 1930s, including notables such as

---

[34] Teller (1972), Tryon (1973), Carey (1978), Sciama (1953).

[35] ten Bruggencate (1948), which originally appeared in the proceedings of the Göttingen Academy in 1945.

[36] Jordan (1938). As mentioned, in the same year Solomon suggested a similar idea.

Wolfgang Pauli, Enrico Fermi and Niels Bohr. According to Pauli, "present-day classical field theories, including the relativistic theory of gravitation, do not give a satisfactory interpretation of the essentially *positive* character of the constant κ [$=8\pi G/c^2$], which is responsible for the fact that gravitation manifests itself as an attraction and not a repulsion of gravitating masses." He continued:

> Such an interpretation could consist only in the reduction of the constant κ to the *square* of another constant of nature. This suggests looking for phenomena in which the square root of the constant κ plays a part. While hitherto it has been regarded as almost certain that gravitational phenomena play practically no part in nuclear physics, it now appears that the possibility that the phenomena of β-radioactivity might be connected with the square root of κ can no longer be rejected out of hand.[37]

In the case of radiometric dating by means of beta decay (as in the Rb-Sr and K-Ar methods) Jordan hypothesized that beta-active elements did not follow the ordinary Rutherford–Soddy exponential law

$$\frac{dN}{dt} = -\lambda N(t),$$

where λ is the decay constant. Instead it might follow a law of the form

$$\frac{dN}{dt} = -\lambda^* N(t)\sqrt{t}.$$

Here λ^* is a true constant, meaning that the measured mean lifetime varies as $t^{-\frac{1}{2}}$. With this hypothesis he hoped to bring the age of the oldest rocks into agreement with his cosmology based on the $G(t)$ hypothesis. In a later paper co-authored by Fritz Houtermans the hypothesis was corrected to

$$\frac{dN}{dt} = -\lambda^* N(t)\frac{1}{\sqrt{t}} \quad \text{or} \quad N(t) = N_0 \exp\left(-2\lambda^* \sqrt{t}\right).$$

The time variation was assumed to be valid for positive and negative beta decay and for K-capture, but not for alpha decay.[38] The German physicist Helmut Hönl, a former student of Arnold Sommerfeld, agreed that the Jordan–Houtermans approach or some other application of Dirac's ideas might be the best way to solve the grave time scale difficulty.[39]

---

[37] Pauli (1936), p. 76; see also Jordan (1944), p. 186.

[38] Houtermans and Jordan (1946), Jordan (1947a), p. 20. Together with the British physicist Robert Atkinson, in 1929 Houtermans pioneered quantum-mechanical nuclear astrophysics. See Kragh (1996), pp. 85–87. After World War II he specialized in geophysics and meteoritics, including radiometric dating methods. In 1953 he estimated the age of the Earth to be 4.5 billion years. For Houtermans' work and eventful life, see Amaldi (2012).

[39] Hönl (1949). See also Dehm (1949).

As late as 1969 Jordan considered it a possibility that the Fermi constant governing beta decay might vary in time in a manner similar to the variation of gravity. He suspected that the age of the Earth as determined by beta radioactivity would be greater than the age as inferred by alpha radioactivity.[40] The hypothesis was taken up at a conference at the Massachusetts Institute of Technology in June 1957, where Houtermans and other participants—including geologists as well as physicists—discussed the ideas of Dirac, Jordan, Dicke and other physicists that some of the constants of nature might vary with time.[41] The question of a variation in time of beta decay was investigated empirically by two Canadian physicists, E. Kanasewich and J. Savage from the University of British Columbia, Vancouver, who concluded that the "Rutherford theory" of no time-variation agreed better with data than the Dirac theory.[42]

Jordan was not the only physicist of repute to consider G-dependent decay times of nuclear particles and nor was he the only one to move into the foreign land of geophysics in the post-World War II era. The British physicist Patrick Blackett, a Nobel laureate of 1948, had no previous training in geophysics. He entered the field as an amateur and mostly learned it himself through a close reading of Holmes' *Principles of Physical Geology*. In the late 1940s Blackett proposed a fundamental theory of terrestrial magnetism which aimed at connecting gravitation and magnetism. He suggested that any rotating body, as a result of its rotation, possessed a magnetic field. Although the "Blackett effect" was never confirmed, it led him to design a sensitive magnetometer for paleomagnetic investigations. His extensive studies of geomagnetism soon turned him into a supporter of continental drift.[43]

Some of Blackett's ideas were clearly inspired by the LNH and similar cosmophysical speculations of a numerological kind. For example, he found that the magnetic moments of the Earth and the Sun were nearly proportional to their angular momenta and that the constant of proportionality could be related to the inverse of Dirac's large number in the form $e^2/m^2G \cong 10^{42}$. Whereas his suggestion of gravitational magnetism was dismissed by most theoretical physicists, in letters from about 1950 Jordan expressed interest in it.[44] Numerology in the style of Dirac and Jordan was not new to Blackett. In 1939 he speculated that the mean life τ_0 of the "mesotron" (or meson, now the muon) might depend on the gravitational constant. Blackett's expression for the dependency can be written as

---

[40] Jordan (1969b), pp. 254–255. See also Jordan et al. (1964), p. 505.

[41] See Aldrich et al. (1958). This was possibly the first scientific conference ever specifically devoted to the relationship between cosmology and geology.

[42] Kanasewich and Savage (1963).

[43] Blackett's important work in geomagnetism and other parts of geophysics is considered in Frankel (2012b) and Nye (2004).

[44] According to Nye (2004), p. 17. See also Jordan (1955), pp. 271–272.

Fig. 2.5 Physicists in front of the Niels Bohr Institute, Copenhagen, commemorating Niels Bohr and the fiftieth anniversary of his atomic model. Jordan (*second row to the left*) was at the time busy developing his theory of the expanding Earth, Dirac (*first row*) was beginning to reconsider his varying-gravity theory, and Blackett (*first row*) focused on geomagnetism and continental drift. On the *first row, from the left*: A. Pais, F. Bloch, F. Hund, W. Houston, C. Møller, D. Dennison, I. Rabi, V. Weisskopf, Aa. Bohr, P. Dirac, O. Frisch, O. Klein, W. Heisenberg, P. Blackett, R. Courant, and A. Rubinowicz. On the *third row, third and fourth from the right*: J. Wheeler, C. Weizsäcker. Credit: Niels Bohr Archive, Copenhagen

$$\tau_0 = \frac{e^3}{\mu^2 c^3 \sqrt{G}},$$

where μ denotes the mass of the mesotron, about 200 electron masses.[45] "It was shown," Blackett commented a few years later, "that the mean life of the meson at rest probably depended on the gravitational constant, and so, through general relativity theory, on the total mass of the universe."[46] He thought that the mean life might vary as the square root of the age of the universe. The idea was taken up by Jordan in his proposal of a possible time-variation of the decay constant of beta radioactivity. Whereas Blackett did not assume Dirac's $G(t)$ hypothesis, Jordan did, which caused him to suggest that the lifetime of beta-emitters varied proportionally to the square root of the age of the universe.[47] Since the mid-1950s Blackett and Jordan increasingly focused on geophysics, the first turning into an advocate of continental drift and the second of the expanding Earth (Fig. 2.5).

---

[45] Blackett (1939).

[46] Blackett (1941), p. 213.

[47] Jordan (1944).

2.4 "A Landmark in Human Thought"

The general idea of time-varying laws of physics, including a gravitational constant varying in time, had been anticipated by Milne, Professor of Mathematics at the University of Oxford, in his theory of so-called kinematic cosmology.[48] This theory rested on a penetrating analysis of the concept of time which led him to conclude that nature can be described by at least two different time parameters. Optical and atomic phenomena would run according to "kinematic time" (t), whereas mechanical phenomena such as pendulum clocks, planetary motions and the rotation of the Earth kept "dynamical time" (τ). Milne found that the two time scales were connected by

$$\tau = t_0 \log\left(\frac{t}{t_0}\right) + t_0 \quad \text{or} \quad \frac{d\tau}{t_0} = \frac{dt}{t}.$$

The constant t_0 can be interpreted as the present epoch, making the two times equal at present. On the t-scale the gravitational constant varied as $G \sim t$ and the universe was expanding from a singular point in space; but on the τ-scale G would remain constant and the universe be static, with its history stretching infinitely backwards to $\tau = -\infty$. Likewise, although the age of the Earth was approximately 3×10^9 years as determined from radioactive decay (t time), in the past the dynamical year was shorter than the kinematic year and consequently the Earth was much older as measured in the dynamical τ-time.

Based upon kinematic time Milne reasoned that the constant of gravitation would increase with the epoch according to

$$G = \frac{c^3}{M_0} t.$$

Milne referred to M_0 as the fictitious or apparent mass of the universe, namely, the extrapolated mass around an observer from $r = 0$ to $r = ct$. With $t = T_H$ he found $M = 2.55 \times 10^{52}$ kg, which corresponds nicely to the number of nucleons in the apparent universe of the same order as the Eddington–Dirac number 10^{79} (1 kg $\cong 6 \times 10^{26}$ nucleon masses, meaning $M = 1.5 \times 10^{79}$ nucleon masses). However, Milne's real universe was uniform and infinite, hence containing an infinite number of particles.

The relation $G \sim t$ had the advantage that shortly after $t = 0$, when the galaxies were closely packed, there would be no gravitational pull to brake the rapid expansion. With increasing time G would grow, but now the galaxies would be so far apart that gravitation could be ignored. As Milne pointed out, in Dirac's theory the situation was just the opposite. In a paper of 1938, where he extended his kinematic relativity theory to cover also electromagnetic phenomena, Milne

---

[48] Milne (1935). The theory is analysed in Cohen (1949–1950) and North (1965), pp. 149–185.

considered Dirac's LNH. However, he concluded that there was no room for G decreasing inversely with the cosmic epoch within his own system.[49]

The relative change of G in Milne's system would be about $dG/Gdt = 5 \times 10^{-10}$ per year, but the rate was of no great significance to Milne, who placed all of his results, including $G \sim t$, in a conventionalist perspective. He emphasized that the relation did not imply that local gravitation, as in the solar system or on the surface of the Earth, increases in strength. In fact, he did not consider $G \sim t$ to be subject to experimental test at all. Frederick L. Arnot, a lecturer of physics at St. Andrews University in Scotland, developed a cosmological theory along the lines of Milne and Dirac.[50] On the basis of Milne's two time-scales and "without any appeal to general relativity" he showed that

$$GM = kc^3t.$$

The quantity k is a dimensionless constant of the order of unity. For the total number of particles he obtained, like Dirac in 1937,

$$N = \frac{1}{2}t^2,$$

where t is expressed in atomic units. Contrary to Dirac and Milne, in Arnot's system the universe was not expanding in either of the two time-scales. To account for Hubble's redshift-distance relation he proposed that the speed of light varied in time.

Despite the obvious differences between the cosmological systems of Dirac and Milne, the latter's ideas had a considerable impact on Dirac's thinking. In an interview with the physicist and author David Peat from the early 1970s, Dirac said: "One should give Milne the credit for having the insight of thinking that perhaps the gravitational constant is not really constant at all. Nobody else had questioned that previously."[51] At a conference in Tallahassee in 1975 Dirac similarly praised Milne's distinction between the atomic and the mechanical time scale. "In fact," he said, "it really started me off on this whole line of work."[52]

The British evolutionary biologist John B. S. Haldane at University College London was fascinated by Milne's ideas. In two papers of 1945 he developed this "landmark in human thought," as he called Milne's theory, into a speculative cosmological theory. The result was nothing less than "a quantum theory of the origin of the solar system."[53] There is no need to deal in detail with Haldane's

---

[49] Milne (1938).

[50] Arnot (1938). See also the more elaborate theory put forward in Arnot (1941).

[51] Online as http://www.fdavidpeat.com/interviews/dirac.htm.The interview is published in Buckley and Peat (1979).

[52] Dirac (1978c), p. 8.

[53] Haldane (1945a, b). See also Barrow and Tipler (1986), pp. 244–245 and Kragh (2004), pp. 221–224.

admittedly amateurish and "wildly speculative" theory (his own expression) except noting that it resulted in a new picture of the history of the Earth. The radioactive decay law translated from kinematic to dynamical time by means of $d\tau/t_0 = dt/t$ yields

$$\frac{dN}{d\tau} = -\lambda N(t)\frac{t}{t_0}.$$

It follows that the rate of decay will first increase and only subsequently, after having passed a maximum at $t = 1/\lambda$, decrease. This, Haldane suggested, led to a new picture of the heat economy of the Earth that differed from the one obtained by John Joly, Arthur Holmes and most other physical geologists: "We reach the surprising conclusion," Haldane wrote, "that the heat production per year by radioactivity in the earth's crust is not diminishing, as has been assumed so far, but is increasing and will increase for some 1000 million years."[54] Surprising indeed.

Haldane did his best to interest astronomers and geologists in his theory, but his attempts were completely unsuccessful. Although Milne endorsed it enthusiastically, it was ignored by the geological community. Thornton Page, an American astronomer and former student of Milne, referred to Haldane's theory in a popular review article on the origin of the Earth. Although Page found the theory to be interesting, he also characterized it as "the most bizarre suggestion of all in this field already rich in speculation."[55] Bizarre it was, and soon forgotten.[56] It was Dirac's G (t) theory that eventually established contact between cosmology and geology, and not the Milne–Haldane theory. In his two publications of 1945 Haldane ignored Dirac's theory of varying gravity and stuck to Milne's.

2.5   Paleoclimatology Enters Cosmology

The Hungarian-American nuclear physicist Edward Teller was in 1935–1942 a colleague of Gamow at the George Washington University. He shared Gamow's interest in astrophysics and in 1939 the two collaborated on a theory of galaxy formation within the context of the expanding universe.[57] After wartime work in the Manhattan Project Teller took up a position as professor at the University of Chicago. In 1947 he participated in the tenth Washington Conference on

---

[54] Haldane (1945b), p. 140.

[55] Page (1948), p. 23.

[56] Haldane's theory was critically reviewed by Lemaître and also by the philosopher Robert Cohen. See references in Kragh (2004), p. 225. See also Stanley-Jones (1949). Today the theory is forgotten, and perhaps justly so. It is briefly mentioned in Brush (2001), p. 164.

[57] Gamow and Teller (1939).

Theoretical Physics, the topic of which was "Gravitation and Electromagnetism," and during the meeting he discussed Dirac's cosmological theory with Gamow and the Princeton astrophysicist Martin Schwarzschild.[58] The following year Teller published a brief paper in *Physical Review* in which he argued that the $G(t)$ hypothesis led to consequences in conflict with established paleontological knowledge.[59] His argument was of the retrodictive rather than predictive kind, in so far that he suggested a scenario of what the Earth would have looked like in the past if certain conditions—such as the constant of gravitation—were altered. He then compared this hypothetical past, or rather the present consequences of this past, with the actual Earth. For more than two decades Teller's paper played an important role in the interplay of geology and cosmology, but today it is not well known. Teller seems himself to have considered it rather unimportant, a mere parenthesis in his scientific career.[60]

Teller's argument assumed that the temperature of the Earth's surface depends directly on the energy flux received from the Sun. He did not take into account cloud formation or other meteorological factors. From astrophysical theory he assumed the solar luminosity to vary as

$$L \sim G^7 M^5,$$

where M is the mass of the Sun. The temperature at the Earth's surface, at a distance r from the Sun, follows from the Stefan–Boltzmann law in the form

$$\frac{L}{4\pi r^2} = \sigma T^4 \text{ or } T \sim \left(L/r^2\right)^{1/4}.$$

Classical mechanics yields

$$r^2 v^2 = GMr,$$

where v is the orbital velocity of the Earth. Teller noticed that this quantity will remain constant even if G varies in time, which is a consequence of angular momentum conservation ($L = rmv \sim rv$). It follows that the radius of the Earth's orbit varies as

$$r \sim \frac{1}{GM}.$$

---

[58] Martin Schwarzschild (1912–1997) was the son of the famous German astronomer Karl Schwarzschild (1873–1916). After studies in mathematics and astronomy in Göttingen and Berlin, in 1937 he moved to the United States where he obtained a position at Columbia University and in 1942 became a U.S. citizen. In 1947 he accepted a position at Princeton University.

[59] Teller (1948).

[60] In his extensive memoirs Teller did not even mention his 1948 excursion into cosmology and paleoclimatology. See Teller and Shoolery (2001).

With $G \sim 1/t$ the result becomes

$$T \sim M^{7/4} t^{-9/4}.$$

Teller considered both of the variations Dirac had proposed in 1937, either a decreasing G alone or combined with an increasing M. Assuming $M = $ constant the result is $T \sim t^{-9/4}$, meaning that T depends sensitively on the age of the Earth. The temperature of the Earth's surface in the past will be related to its present temperature T_0 by

$$T = T_0 \left(\frac{t_0}{t}\right)^{9/4}.$$

Here t_0 denotes the present epoch or roughly the Hubble time. Teller concluded that at a time 200–300 million years ago, "We are led to expect a [surface] temperature near the boiling point of water." The gravitational constant would at the time have been approximately 10 % greater than its present value. On Dirac's original assumption $N(t) \sim t^2$ and consequently $M \sim t^2$, the result is a slowly increasing temperature over geological time following

$$T \sim t^{1/4}.$$

Teller did not actually give this result but only illustrated the cool ancient Earth with an example, stating that in the same time interval the temperature would have been 12 % lower than today. "This would bring the average temperature on the earth below the freezing point," he wrote. The actual values of Teller's two examples of past temperatures are 79 and $-15\,°C$.

Since geology and paleobiology showed "ample evidence of life on our planet at this time," Teller felt justified in concluding that Dirac's hypothesis was wrong or seriously inadequate. However, he realized that the argument was perhaps oversimplified and that the hypothesis might in some way escape the conclusion. "Our present discussion cannot disprove completely the suggestion of Dirac," Teller admitted. "The suggestion is, because of the nature of the subject matter, vague and difficult to disprove." Indeed, it would take another four decades before the $G(t)$ hypothesis was definitely proven wrong.

Despite Teller's cautionary remark his paper was often cited, with or without proper reference, as proof of the incorrectness of Dirac's cosmology based on the $G(t)$ hypothesis. For example, without mentioning Teller by name Paul Couderc stated that Dirac's theory "involves a variation of the terrestrial temperature during the past 500 million years, which geological observations utterly denies."[61] Other physicists and astronomers followed troop.[62] However, as early as 1950 Dirk ter

---

[61] Couderc (1952), p. 98.

[62] E.g., Gamow (1949), p. 21, and Omer (1949), p. 166. See also Schatzmann (1966), p. 219, originally published 1957 as *Origine et Évolution des Mondes*.

Haar, a British-Dutch physicist at the University of St. Andrews, pointed out that Teller's argument was inconclusive because it rested on a number of *ceteris paribus* assumptions. One of them was the assumption that the opacity of the Sun remained the same and another was the disregard of cloud formation in the atmosphere. Ter Haar argued that if heavy clouds were formed all over the Earth, the average temperature would be considerably lowered.[63]

Jordan agreed with ter Haar and amplified his arguments against the "Gamow–Teller objection" that the Sun's heat would have caused "the trilobites in the oceans to boil," as he phrased it.[64] Presenting Teller's line of reasoning in a slightly different way he expressed the luminosity of the Sun (of radius R) as

$$L \sim M^{11/2} R^{-1/2} G^{15/2}.$$

Jordan deduced that the solar constant S would depend very strongly on the gravitational constant, namely as $S \sim G^{10}$, although he admitted that this was only a rough approximation. The dependence of S on G might well be less pronounced.[65] In his later work Jordan found that, with S_0 a constant and $dG/Gdt = 10^{-9}$ per year, the time-dependence of S would approximately follow

$$S \cong S_0\left(1 - 10^{-8}t\right),$$

where t is measured in years.[66] This meant that the solar constant in the Carboniferous was approximately 50 % greater than today. According to Jordan, an increase in the Sun's luminosity would first of all result in a large number of heavy clouds, perhaps turning the Earth's atmosphere into something resembling the atmosphere of Venus. He suggested that there was convincing geological and paleoclimatic evidence that the Earth had in fact been covered by clouds in the Carboniferous age and part of the Permian age.

Teller's argument against Dirac's hypothesis of a decreasing gravitational constant was well known and often referred to, if more by physicists and astronomers than by earth scientists. In the mid-1960s it was developed and sharpened, first by the Princeton astronomers Philip Pochoda and Martin Schwarzschild and then independently by Gamow and the Estonian-Irish astronomer Ernst Öpik. By that time the time scale of the universe had been significantly revised in the sense that the Hubble constant turned out to be much smaller than previously thought.[67] The accepted age of the universe had now increased to between 10 and 15 billion years,

---

[63] ter Haar (1950), p. 131.

[64] Jordan (1955), p. 235.

[65] Jordan (1962b), p. 285. See also Dyson (1972), according to whom $S \sim G^{9.7}$ or $S \sim t^{-9.7}$. As mentioned in Sect. 2.3, ten Bruggencate (1948) was probably the first to consider the effect of the $G(t)$ hypothesis on the Sun's luminosity.

[66] Jordan (1964), p. 114.

[67] On the change in the cosmic time scale, see Kragh (1996), pp. 271–276.

which largely—if not completely—removed the time scale difficulty that until then had plagued most finite-age cosmological models.

From about 1960 to 1967 Dirac and Gamow exchanged a series of letters in which they discussed Dirac's cosmological hypothesis and its astronomical and terrestrial consequences. In a letter to Gamow of 10 January 1961, Dirac wrote:

> It was a difficulty with my varying gravitational constant that the time scale appeared too short, but I always believed the idea was essentially correct. Now that the difficulty is removed, of course I believe more than ever. The astronomers now put the age of the universe at about 12×10^9 years, and some even think that it may have to be increased to 20×10^9 years, so that gives us plenty of time.[68]

In reality, even the extended time scale did not give "plenty of time." The values cited by Dirac were either Hubble times or ages of the universe on the assumption of an Einstein–de Sitter universe. In 1958 Allan Sandage at the Palomar Observatory estimated 13×10^9 years as an upper limit for a universe of the Einstein–de Sitter type. As he pointed out, "there is no reason to discard exploding world models on the evidence of inadequate time scale alone."[69] However, the quoted value corresponds to 6.5×10^9 years for the Dirac model and while this value was greater than the age of the Earth—at the time known to be 4.5×10^9 years—it was smaller than the age of the oldest stars.

The extension of the time scale also helped Dirac's theory by weakening Teller's objection. Because, with an age of the universe of 12×10^9 years the temperature of the Earth in the Cambrian era would only be about 45 °C and thus no longer rule out life on Earth. As noted by Pochoda and Schwarzschild, "with the newer, rather long estimates for the age of the Universe, the time elapsed since the Pre-Cambrian appears only a rather modest fraction of the total time scale, so that the excess of G in the Pre-Cambrian over its present value and the consequence excess in the solar luminosity would be quite small."[70] The two Princeton astronomers were mainly concerned with the history of the Sun, not with the history of the Earth. For flat-space cosmological models they computed numerically the evolution of solar models in which $G(t)$ varied as

$$G = G_0 \left(\frac{T}{t}\right)^n,$$

where T is the age of the universe and G_0 represents the present value of G. In the case of Dirac's hypothesis with $n = 1$, they showed that it led to a luminosity of the Sun in its early stages of evolution nearly five times its present luminosity. As a consequence, the Sun would no longer be a main sequence star. To make $n = 1$ agree with the observed Sun, they had to assume $T \geq 15 \times 10^9$ years. Pochoda and Schwarzschild also made calculations in the case of $n = 0.2$, as given by the Brans–

---

[68] Quoted in Kragh (1990), p. 237. For the Dirac–Gamow correspondence, see Kragh (1991).

[69] Sandage (1958), p. 525.

[70] Pochoda and Schwarzschild (1964), p. 587.

Dicke theory (to be considered below). With this mild variation they found results that only differed insignificantly from the constant G corresponding to the standard case $n = 0$ valid for relativistic models. However, according to the astrophysicists G. Shahiv and John Bahcall the Brans–Dicke theory led to problems with the Sun's neutrino flux.[71]

In view of the astronomers' estimates of the Hubble time Pochoda and Scharzschild considered the condition $T \geq 15 \times 10^9$ years to be "uncomfortable." With an age of the Earth of approximately 5 billion years, G at the time of the Earth's formation would have been approximately $1.5\,G_0$, whereas $T = 10 \times 10^9$ years led to $G \cong 2\,G_0$. Using a different method of calculation Gamow similarly showed that on Dirac's hypothesis the Sun could not have shone for more than 2 billion years.[72] It would have burned all its hydrogen and subsequently turned into a red giant.

While Pochoda and Schwarzschild had little to say about the Earth, Öpik was greatly interested in the earth sciences and their connections to astronomy. In 1950 he had proposed a theory of the ice ages and over the next two decades he continued to work on this problem and related problems in paleoclimatology. In a review article of 1965 on climatic changes in "cosmic perspective" he drew on a wealth of geological evidence to explain the climatic balance of the Earth. Characteristically, the article was written at the request of a leading geologist, the Australian Rhodes Fairbridge, who was an early expert on climate change and known in particular for his hypothesis of regular changes in the ocean levels over long time scales. In a footnote to his paper of 1965, Öpik acknowledged Fairbridge's role, adding that the paper was written in 1961, "but its publication was delayed by unfavourable coincidences."[73]

In relation to Teller's paper of 1948, Öpik considered the terrestrial consequences of a varying gravitational constant by calculating the relative insolation (solar radiation energy per unit area) of the Earth on the assumption of Dirac's hypothesis. A comparison of the calculated insolation in the past and the one estimated from the geological record told him that the two were incompatible, meaning that Dirac's $G(t)$ hypothesis must be wrong. "Even the scanty geological data are sufficient to cause rejection of a bold cosmological idea," he concluded.[74] Jordan emphatically disagreed.

Öpik was himself in favour of another bold cosmological idea, namely the eternally oscillating universe and it is possible that his preference for this kind of

---

[71] Shahiv and Bahcall (1969). See also Sect. 2.7.

[72] Gamow (1967c). Other problems in solar nucleosynthesis based on a decreasing gravitational constant were pointed out in Ezer and Cameron (1966).

[73] See also Dicke (1962a), which refers to Öpik's paper of 1961 as "privately circulated." Fairbridge was at the time a supporter of the hypothesis of a slowly expanding Earth, which was sometimes justified in terms of a decreasing gravitational constant. However, Fairbridge did not accept the Dirac–Jordan hypothesis of $G(t)$. We shall return to Fairbridge in Sect. 3.6.

[74] Öpik (1965), p. 292.

universe coloured his attitude to Dirac's hypothesis.[75] According to Dirac, the LNH and hence also $G(t)$ was incompatible with oscillating or cyclic models of the universe. In the latter models the curvature of space must be positive, whereas the LNH leads to a flat space. As another reason for dismissing oscillating universe models Dirac mentioned that such models involve a large number, namely the maximum radius of the universe, which does not fit with the LNH.

The nuclear-physical theory of the early universe that Gamow and his collaborators Ralph Alpher and Robert Herman developed during the period 1946–1956 was the first big-bang cosmology in the modern sense of the term. Supervised by Gamow, Alpher completed in the summer of 1948 his Ph.D. thesis on element formation in the early universe, the first of its kind. He not only mentioned Dirac's LNH and its consequence in the form of the $G(t)$ hypothesis, but also Teller's very recent argument against it.[76] However, neither he nor Gamow accepted Dirac's reasoning. The Gamow–Alpher–Herman theory was solidly based on Einstein's field equations, which Gamow always used in the standard version given by the Friedmann equations and simply took for granted. He consequently disregarded alternative theories such as Dirac's $G(t)$ theory and also Milne's cosmology and the new steady-state theory. Although Gamow found Dirac's LNH to be fascinating and perhaps even true, he denied its purported consequence in the form of a decreasing gravitational constant.

When Gamow and most other physicists denied the validity of Dirac's cosmology it was not only because of its empirical consequences but also because of its conflict with general relativity, according to which G must be constant. But must it? Not according to C. Gilbert, a mathematical physicist at King's College, the University of Newcastle. Gilbert used principles allegedly based on the theory of general relativity to derive a cosmological model similar to Dirac's, including a gravitational constant varying as $G \sim 1/t$.[77] To derive this result he introduced elements from Milne's cosmological system into general relativity. Gilbert's model was characterized by a deceleration parameter $q_0 = \frac{1}{2}$, mean density of matter $\rho = 4.8 \times 10^{-29}$ g cm^{-3}, and an age of the universe equal to 4×10^9 years, corresponding to $H_0 \cong 160$ km s^{-1} Mpc^{-1}. By 1956 this was not only a suspiciously low age, it also conflicted with the authoritative age of the Earth of $(4.55 \pm 0.07) \times 10^9$ years which the American geochemist Clair Patterson published the same year and since then has stood the test of time.

In a later communication Gilbert admitted that his age of the universe was too small.[78] In any case, his theory did not succeed in replacing the standard view that there is no room for a changing G within Einstein's theory of general relativity. Mainstream physicists maintained that an amalgamation of $G(t)$ and general relativity cannot be accomplished consistently. If G is not constant, energy-momentum

---

[75] Öpik (1956).

[76] According to Peebles et al. (2009), p. 38.

[77] Gilbert (1956, 1957). See also Wesson (1978), pp. 22–26.

[78] Gilbert (1961).

conservation cannot be satisfied. As we shall see in Chap. 3, outside mainstream physics Gilbert's claim continued to play a role.

As mentioned above, Gamow was involved in Teller's original argument against Dirac's hypothesis of $G(t)$. Gamow repeated the argument in a paper of 1949 and also in a later popular book.[79] In discussions and correspondence with Dirac from the 1960s he maintained and further developed what Jordan called the Gamow–Teller objection. Then, in a series of papers from 1967, he made public what he had discussed privately with Dirac. Gamow admitted that the longer time scale changed the details of Teller's old argument, but not that it invalidated it. As he pointed out, palaeontologists had recently found fossils of primitive life as old as 3 billion years. "Even though Teller's argument makes life safe for inhabitants of the Cambrian ocean, it certainly threatens the life of organisms living a few eons [billion years] ago."[80] Having discussed the additional problem of the increased solar luminosity in the past, he concluded that "the possibility of the change of gravity in inverse proportion to the age of the universe has been completely ruled out."

While Gamow insisted that the varying-G hypothesis was plain wrong, he wanted to retain the LNH, and for this purpose he suggested that the elementary charge e might instead increase in time according to

$$e^2 \sim t$$

This hypothesis, contrary to Dirac's, did not lead to obvious disagreements with either paleontological or cosmological evidence.[81] Gamow's varying-e hypothesis was however contradicted by new measurements of the fine-structure constant ($2\pi e^2/hc$, where h is Planck's constant) in distant quasars. As soon as Gamow became aware of the contradiction, he retracted the hypothesis.[82] All later measurements have shown that the elementary charge is a true constant of nature. In this connection it is worth noting that late in his life Dirac suggested that Planck's constant h varied with the epoch. Since Dirac believed that the fine structure constant was a true constant, Gamow's relation $e^2 \sim t$ followed.[83]

Thirty-five years after having proposed his cosmological theory Dirac finally returned to it. In a volume devoted to the commemoration of Gamow (who died in 1968) Dirac and Teller commented on the current state of varying-G cosmology. Teller essentially restated his old argument against a decreasing G, except that he

---

[79] Gamow (1949, 1962), pp. 139–141.

[80] Gamow (1967a), p. 759. It should be noted that the geological periods and the times in the past associated with them have changed considerably during the twentieth century. For example, the "Cambrian" did not have a fixed meaning across the century. Gamow apparently associated the Cambrian with Teller's period 200–300 million years ago, but according to Holmes' revised time scale of 1960 the Cambrian ended about 500 million years ago. The presently accepted time for the end of the Cambrian period and the beginning of the Ordovician is 485 million years ago.

[81] Gamow (1967b, c). See also Kramarovskii and Chechev (1971).

[82] Kragh (1991), Wesson (1980), p. 45.

[83] Dirac (1982), p. 88.

now found the consequences for astrophysics to be more decisive than those relating to the past climate of the Earth.[84] In his brief essay, Dirac repeated what he had said privately to Gamow, namely, that Teller's and others arguments were not really a fatal blow to his hypothesis. It could, he vaguely suggested, be rescued by making certain assumptions.[85] What these assumptions were he first spelled out in a lecture to the Pontifical Academy of Sciences on 13 April 1972.[86] By that time he was ready to reintroduce the idea of spontaneous creation of matter in the form of the $N \sim t^2$ hypothesis that he had abandoned in 1938. The year 1972 also saw the publication of a book dedicated to Dirac's seventieth birthday in which the distinguished theoretical physicist Freeman Dyson systematically and critically reviewed Dirac-inspired ideas of the variation of fundamental constants.[87] Dyson's essay was instrumental in reviving interest in the subject.

Dirac subsequently developed his cosmological theory in several versions, most of which made use of the varying-G hypothesis $G \sim 1/t$ and also the creation hypothesis $N \sim t^2$.[88] None of these versions won much support, although they attracted considerable interest and inspired many comments. In his publications on cosmology, confined to the periods 1937–1938 and 1972–1982, Dirac paid little attention to the history of the Earth, a subject he never took seriously and only referred to casually. But a few other cosmologists did take the subject seriously, in particular Jordan and Dicke to whose theories of a cosmology–geophysics connection based on the assumption of a varying G we now turn.

2.6 Offspring of Scalar–Tensor Gravitation Theory

In Einstein's general theory of relativity, gravitation is fully described by the geometry of space-time as given by the metrical tensor $g_{\mu\nu}$. According to the class of "scalar–tensor" theories developed in the 1950s and 1960s a scalar field φ must be added to the tensor equations to account for the measurable value of the gravitational constant. The value of the new φ field, and hence the gravitational constant, depends on the point in space-time, $\varphi = \varphi(x, y, z, t)$.

The scalar–tensor formalism became well known only after Robert Dicke and Carl Brans published a paper on it in 1961, but most of the formalism had been established earlier by Jordan and his collaborators. In fact, elements of the scalar–

---

[84] Teller (1972).

[85] Dirac (1972).

[86] Dirac (1973d).

[87] Dyson (1972).

[88] Wesson (1978), pp. 13–19, Kragh (1990), pp. 239–245. More about Dirac's later theories follows in Sect. 4.2.

tensor formalism can be found even earlier.[89] The kind of theory is often known under the names of Brans and Dicke alone, but Jordan–Brans–Dicke (JBD) is also used.[90] Several other physicists were involved in the development of this kind of theory. As Brans recently noted, "if a list of names were to be used for people who independently proposed ST [scalar–tensor] modifications of standard Einstein theory, the resulting compound title would be extravagantly unwieldy."[91] Contrary to most other physicists, Dicke was always careful in acknowledging Jordan's priority. What matters in the present context is that the theories of this class include a gravitational constant that varies in time if not necessarily in Dirac's sense.

It is worth noting that much of the work done by Jordan's group in Hamburg was collaborative and not always with Jordan playing a leading or even an active part. According to George Ellis, a leading mathematical cosmologist, "Jordan's name was on . . . the papers, although he in fact did not take part in writing them: his name was there simply because he was the head of the group from which they came."[92] This was also the opinion of Jordan's collaborator Engelbert Schücking: "Jordan appeared often as co-author, but I doubt whether he contributed much more than suggestions in style, like never to start a sentence with a formula."[93] Some of the research done by Jordan and his group was supported by the Aeronautical Research Laboratories (ARL), an institution under the U.S. Air Force. In the 1960s Jordan wrote two ARL reports on his extension of general relativity and the $G(t)$ hypothesis.[94]

Jordan originally formulated his theory within the framework of the five-dimensional, so-called Kaluza–Klein unified theory going back to the 1920s.[95] He was not the only one who in the 1940s tried in this way to make place for a scalar field corresponding to a varying gravitational constant. So did Einstein and his collaborator, the German-American mathematical physicist Peter Bergmann. In a paper of 1948 the latter recalled that in the spring of 1946 Wolfgang Pauli (who at the time resided in the United States) turned over to him the proofs of a paper that Jordan had written on gravitation theory with a varying gravitational constant. The

---

[89] The first scalar–tensor gravitational theory was published in 1941 by the Swiss mathematician Willy Scherrer at the University of Bern. For the origins and early development of scalar–tensor theories, see Goenner (2012).

[90] Dehnen and Hönl (1968, 1969) may have been the first to use "Jordan–Brans–Dicke" and also the abbreviation "JBD." Occasionally physicists speak of "Dicke–Brans–Jordan" theory (DBJ) and other permutations are also in use.

[91] Brans (2014).

[92] Ellis (2009), p. 2180.

[93] Schücking (2000), p. vi.

[94] The ARL reports are listed in Goldberg (1992). One of Jordan's reports contained a detailed review of geophysical and astronomical aspects related to Dirac's hypothesis. See Jordan (1961b), a copy of which is located at the Niels Bohr Institute, Copenhagen.

[95] Jordan (1948). "Kaluza–Klein" refers to the German mathematician Theodor Kaluza and the Swedish physicist Oskar Klein. While Kaluza's unification comprised gravitation and electrodynamics, Klein's theory also included the quantum domain.

paper was to have appeared in the *Physikalische Zeitschrift* in 1945, but due to the war the journal ceased publication. According to Bergmann:

> In this paper, Jordan attempted to generalize Kaluza's five dimensional unified field theory by retaining g_{55} as a fifteenth field variable. Professor Einstein and the recent author had worked on the same idea several years earlier, but had finally rejected it and not published that abortive attempt. The fact that another worker in this field has proposed the same idea, and independently, is an indication of its inherent plausibility.[96]

Under the impact of work done by his collaborators Günther Ludwig and Claus Müller, Jordan eventually abandoned the five-dimensional theory. Instead he proposed field equations in the usual four dimensions involving a scalar field related to replacing Newton's constant of gravitation.

From the late 1940s onwards Jordan developed his cosmological ideas into an alternative or extended theory of general relativity. At the same time he also became increasingly interested in the connections between cosmology and geophysics, a subject that would dominated much of Jordan's later scientific work and to which we shall return in Sect. 3.2.[97] In the monograph *Schwerkraft und Weltall* (Gravitation and Universe) from 1952 he presented a new basis for his theory of a universe increasing in mass as $M \sim t^2$ and governed by a gravitational force varying inversely with the age of the universe. Jordan's general field equations included two new numbers or dimensionless constant of nature (ζ, η) the values of which were not specified by theory but could only be estimated by comparison with empirical data. On the other hand, there was no room in Jordan's equations for a cosmological constant, that is, $\Lambda = 0$. Jordan argued that $\eta = \pm 1$ and that $\eta = +1$ in order that his theory should comply with the original Dirac hypothesis including matter creation.

For a positively curved space Jordan's theory led to differential equations for the scale factor $R(t)$ that corresponded to the Friedmann equations of ordinary general relativity, only were Jordan's equations more complicated and included more solutions. For $\eta = +1$, $c = 1$ and $\zeta > 2$ he found the linear solution

$$R = \left(\frac{2}{\zeta - 2} \right)^{\!\frac{1}{2}} t.$$

With G_0 and ρ_0 denoting two constants, Jordan derived for the gravitational "constant" G and the mean density of matter ρ that

---

[96] Bergmann (1948), p. 255. The title of Jordan's unpublished paper was "Gravitationstheorie mit veränderlicher Gravitationszahl" (Gravitation theory with variable gravitational constant). Bergmann was sceptical with regard to Jordan's theory because the extra scalar variable caused an *embarrass de richesse*, as he expressed it.

[97] Jordan's enduring interest in geophysics and other aspects of the earth sciences is documented by his many publications on the subject. See the bibliography of Jordan's articles and books in Beiglböck (2007).

$$G = G_0 t^{-1} \quad \text{and} \quad \rho = \rho_0 t^{-1}.$$

In the spherical case the mass of the closed universe varied in time as

$$M = 2\pi^2 \rho_0 \left(\frac{2}{\zeta - 2} \right)^{3/2} t^2.$$

Whereas Dirac only assumed a temporal variation of G, in Jordan's theory G was allowed to vary in both space and time. However, he chose to disregard the spatial variation since it was supposedly unimportant and of no explanatory value.[98]

As to the increase of the mass of the universe according to $M \sim t^2$, Jordan eventually reached the conclusion that the hypothesis was not, after all, tenable.[99] He now declared his earlier belief in matter creation to be "a misunderstanding" and "surely erroneous." Instead he suggested that $M \sim t^2$ was valid only for so-called pre-stellar matter pouring out from pockets of empty space.[100] The final form of Jordan's field equations was indebted to a critical analysis of Markus Fierz at Basel University who pointed out that some of Jordan's equations would lead to atomic spectra depending on space-time. By incorporating the criticism of Fierz and Pauli, in 1959 Jordan proposed a new version of his scalar–tensor theory with mass conservation.[101]

While Jordan's extended gravitation theory only attracted limited attention outside Germany, it was taken up and further developed by several of his colleagues in the new *Bundesrepublik Deutschland*, the German Federal Republic. Among those who examined the theory the young Berlin theoretical physicist Kurt Just was perhaps the most prolific. In a paper of 1955 Just investigated in mathematical detail the cosmological models allowed by Jordan's theory, focusing on the spatially closed model characterized by $G \sim 1/t$ and $R \sim t$.[102] Just mentioned several objections against Jordan's favoured model, one of them being Teller's climatological argument of 1948. Moreover, Just showed that according to Jordan's theory the present radius of curvature would be so small that the positively curved universe should turn up observationally. Yet in the mid-1950s there was no observational evidence indicating cosmic space being positively curved.

The cosmological background radiation discovered in 1965 provided a further argument against Jordan's original theory of gravitation. Helmut Hönl and Heinz Dehnen, two German astrophysicists at the University of Freiburg im Breisgau,

---

[98] Jordan (1971), p. 2.

[99] Jordan (1959a), p. 113.

[100] Jordan (1962c).

[101] Fierz (1956), Goenner (2012). For a lucid summary of Jordan's cosmological theory as developed in the late 1950s, see Heckmann and Schücking (1959). Other aspects of Jordan's theory were dealt with in Brill (1962) and O'Hanlon and Tam (1970).

[102] Just (1955). On Just's early work on Jordan's theory, see Goenner (2012). Just later moved to the United States to become professor of physics at the University of Arizona, Tuczon.

believed that results from astrophysics and cosmology were much better suited to test gravitation theories than the uncertain geophysical results on which Jordan focused.[103] They consequently investigated what the background radiation would look like on the basis of Jordan's theory.[104] In that case the total energy density of the radiation would increase in time and its intensity at different wavelengths would no longer lie on a blackbody-curve at constant temperature. Since these consequences disagreed with measurements, Hönl and Dehnen concluded that the microwave background amounted to a strong empirical argument against a gravitational constant varying as Dirac had proposed.

Accepting the objection raised by Hönl and Dehnen, Jordan now declared his previous equations to be "false." As an alternative he suggested a modification of the field equations that made his theory almost identical with the theory of Brans and Dicke to be considered in Sect. 2.7.[105] In this theory G varies as t^{-n} with $n < 1$. Jordan stressed that although the Brans–Dicke approach differed substantially from his own reasoning, in the end the two approaches led to the same result. This he considered a strong argument in favour of the Jordan–Brans–Dicke theory, which he judged to be "the only possibility of a plausible generalization of Einstein's theory of gravitation."[106] Jordan's new theory still allowed G to decrease in time, but no longer in Dirac's strong version of $G \sim 1/t$. This move did not imply that Jordan abandoned Dirac's hypothesis, but only that he separated it from the gravitational field equations and instead considered it an independent, empirically justified hypothesis. Moreover, although he felt forced to abandon the $M(t)$ hypothesis, he did it with regret. Somehow, he believed, the hypothesis of spontaneous matter creation ought to play a role in the final picture of the universe. Jordan promised to return to the problem, but he never did.

Ideas similar to Jordan's extended theory of general relativity, but developed independently, appeared in an important paper published by the two Princeton physicists Carl Brans and Robert Dicke in 1961.[107] Although at first the paper did not make much of an impact, within a few years it became recognized as a most interesting alternative to standard general relativity. Its continuing appeal is illustrated by bibliometric data: until 1981 it had been cited in about 500 articles and today the cumulative number of citations is more than 2.800 (Fig. 2.6).[108] Brans had been a Ph.D. student of Dicke and the joint paper was an outgrowth of Brans' doctoral thesis from the same year.[109] Still when Brans had almost completed his

---

[103] Dehnen and Hönl (1969), Wesson (1978), pp. 36–37.

[104] Hönl and Dehnen (1968).

[105] Jordan (1968, 1971, p. xv).

[106] Jordan (1969b), p. 253.

[107] Brans and Dicke (1961).

[108] Kaiser (1998).

[109] Brans (1961). See also Brans (1999, 2010) for personal and historical comments on the paper and on scalar–tensor theories in general. Kaiser (1998, 2007) compares the Brans–Dicke theory to the theory of the Higgs field proposed a few years later. Both theories related to the origin of mass, but they belonged to two widely different research traditions.

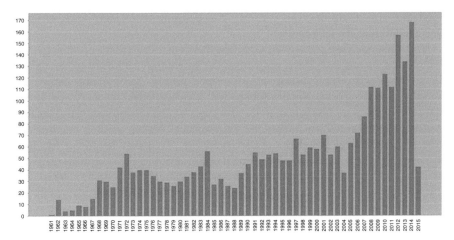

Fig. 2.6 Citations to Brans and Dicke (1961). Credit: Web of Science

thesis he was unaware of Jordan's theory, but (undoubtedly due to Dicke) the final form of it included an extensive discussion of the theory as presented in Jordan's book of 1955. Brans briefly noted that *Schwerkraft* contained a detailed account of the geophysical consequences of Jordan's theory of $G(t)$, but without mentioning which. Nor did the Brans–Dicke paper refer to consequences of a geological or geophysical nature.

2.7   A Machian Approach to Fundamental Physics

What soon became known as the Brans–Dicke theory was based to a large extent on two principles of a general, almost philosophical nature. One of them was Dirac's LNH and the other was Mach's principle according to which (in one of its several versions) the space-time metric is determined by the mass of the universe. In agreement with an earlier idea of Dicke,[110] the two Princeton physicists took the latter principle to imply that the mass M of the visible universe was related to its space curvature radius R by

$$\frac{GM}{Rc^2} \cong 1 \quad \text{or} \quad \frac{1}{G} \cong \frac{M}{Rc^2}.$$

Thus, as the universe expands and M and R change, G changes accordingly. A relation of this kind can also be derived from ordinary Friedmann cosmology, but Brans and Dicke looked upon it as the outcome of individual contributions from celestial bodies at various distances. This alone suggests that G may not be a

---

[110] Dicke (1959a).

fundamental constant but is a quantity which depends on the large-scale structure and evolution of the universe.

The same equation connecting G, M and R was part of Jordan's cosmology, but without associating it to Mach's principle. On the contrary, although Jordan valued Ernst Mach's positivist philosophy of science greatly, he explicitly denied the validity of Mach's principle as a foundation of cosmological theory.[111] Dicke, on the other hand, was fascinated, even obsessed, by the principle of the Viennese philosopher-physicist, which was the philosophical beacon guiding much of his research concerning gravitation and cosmology.[112] In part based on Machian arguments Dicke reached the conclusion that Einstein's theory of general relativity might not be completely correct. Although he admitted that the Einstein theory was elegant, he was "not sure that nature has quite the predilection for an elegant theory that man apparently possesses."[113]

To construct a new theory of gravitation "which is more satisfactory from the standpoint of Mach's principle than general relativity" Brans and Dicke replaced the constant of gravitation with a long-range scalar field $\varphi(x, y, z, t)$ generated by the matter in the universe. The two quantities were inversely proportional, that is, $G \sim \varphi^{-1}$ (contrary to Jordan, who in his early versions had $G \sim \varphi$). This means that a variation of G implies a variation of the scalar field:

$$\frac{1}{\varphi}\frac{d\varphi}{dt} = -\frac{1}{G}\frac{dG}{dt}$$

However, the measured value of G also depended on a dimensionless parameter, ω, the value of which was not given by theory but could only be determined by observation. The relationship between G and ω was stated as

$$G = \left(\frac{2\omega + 4}{2\omega + 3}\right)\varphi^{-1}.$$

Dicke characterized ω as a measure of the fraction of the gravitational force caused by the scalar field. The ω parameter was constant in the sense that it did not depend on the φ field, but later extensions of the Brans–Dicke theory included the possibility of $\omega = \omega(\varphi)$. While Dicke disregarded the cosmological constant, some of the later extensions also included $\Lambda = \Lambda(\varphi)$.

In the lower limit $\omega = 0$ the equations of the Brans–Dicke theory described a model universe in which distance and gravitation varied as

---

[111] Jordan (1955), p. 138.

[112] E.g. Dicke (1964a). Mach's principle has given rise to a very extensive literature, both philosophical and scientific. For a contemporaneous and critical discussion of its use in and relevance for cosmology and theories of gravitation, see for example Reinhardt (1973).

[113] Dicke (1959c), p. 621.

$$R \sim t^{1/3} \quad \text{and} \quad G \sim t^{-1/2}.$$

In the limit $\omega \to \infty$ the equations passed asymptotically over into those of the ordinary Einstein theory of relativity.[114] As in the theories of Dirac and Jordan, gravitation would decrease in time, but at a rate that depended on the value of the ω parameter. For example, in the special case of a flat space with vanishing pressure (what cosmologists call a "dust model"), the rate was given by

$$G = G_0 \left(\frac{t}{T_0}\right)^{-\eta},$$

with

$$\eta = \frac{2}{4 + 3\omega}.$$

T_0 is the age of the universe and the subscript 0 refers to the present era. The variation could also be expressed as

$$\left(\frac{1}{G}\frac{dG}{dt}\right)_0 = -\frac{H_0}{1 + \omega}.$$

For comparison, the rate of decrease of G in Dirac's theory (and in Jordan's) was given by $-3H_0$. "The resulting rate of decrease of the gravitational constant is 1 part in 10^{11} parts per year," Dicke wrote. "With a closed universe this rate of decrease could be as great as 3 parts in 10^{11} per year."[115] In the case of $\omega = 6$ the rate is about twenty times less than according to Dirac's theory and

$$G \sim t^{-0.09}.$$

Dicke further stated the relationship between the age of the universe and the Hubble time T_H to be

$$T_0 = \frac{2 + 2\omega}{4 + 3\omega} T_\mathrm{H}.$$

For $\omega = 6$ this amounts to a value of T_0/T_H that is only marginally smaller than the ratio 2/3 in the Einstein–de Sitter cosmological model. Applying their theory to the anomalous precession of Mercury's perihelion, Brans and Dicke found a value for the perihelion shift that was smaller than Einstein's by a factor given by

---

[114] Brans and Dicke (1961), Dicke (1962a).
[115] Dicke (1962a), p. 657.

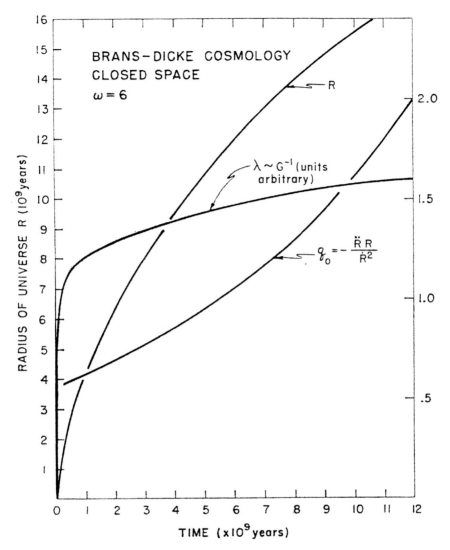

Fig. 2.7 The Brans–Dicke universe. R is the scale factor and q_0 the curvature parameter. The quantity λ is a scalar given by $\lambda = G_0/G(t)$. *Source*: Dicke (1962a), p. 636. Reproduced with the permission of the American Association for the Advancement of Science

$$\frac{4 + 3\omega}{6 + 3\omega}.$$

To secure agreement with observations they concluded that $\omega \geq 6$ and possibly $\omega \cong 6$, which was their preferred value.

Assuming that "the universe expands from a highly condensed state," Brans and Dicke further examined various cosmological models modified by the assumption of a slowly decreasing gravitational constant (Fig. 2.7). While in these models

G varied rapidly in the early stages of the universe, the variation was much slower at later times. In the standard Friedmann theory governed by general relativity, the expansion of the universe of average mass density ρ and zero pressure follows the expression

$$\left(\frac{R'}{R}\right)^2 + \frac{k}{R^2} = \frac{8\pi G\rho}{3}.$$

The speed of light c is taken to be unity, k is the curvature parameter attaining the values 0 or ± 1, and $\Lambda = 0$. In the Brans–Dicke theory the analogous expression is

$$\left(\frac{R'}{R}\right)^2 + \frac{k}{R^2} = \frac{8\pi\rho}{3\varphi} - \frac{\varphi'}{\varphi}\frac{R'}{R} + \frac{\omega}{6}\left(\frac{\varphi'}{\varphi}\right)^2.$$

It is seen that with $\varphi = \text{constant} = 1/G$ the former equation is regained.

Brans and Dicke pointed out that for $\omega \geq 6$ their model for a flat-space universe was practically indistinguishable from the Einstein–de Sitter model based on ordinary general relativity. In another paper Dicke investigated in detail the cosmological and astrophysical consequences of the Brans–Dicke theory, concluding that it was the only theory able to explain the disagreement between the age of the universe and various evolutionary ages of stars and galaxies.[116] In this respect he considered the Brans–Dicke theory to be superior to Dirac's, for on the assumption of $G \sim 1/t$ the age of the Sun and the other main sequence stars would be suspiciously smaller than the usually accepted ages.[117] Contrary to the early versions of Jordan's theory, Brans and Dicke constructed their scalar–tensor theory in such a way that mass–energy was conserved in agreement with standard general relativity.

Many of the ideas of the Brans–Dicke paper of 1961 had been discussed earlier by Dicke, who since the mid-1950s increasingly turned from radiation and quantum physics to studies of gravitational physics, both experimentally and theoretically.[118] He soon became a leading figure in what is known as the "renaissance of general relativity."[119] Dicke's interest in gravitation theory and cosmology, including aspects of geophysics, was first developed during a sabbatical year he spent at Harvard in 1954. Since there was nobody at Harvard to discuss the subjects with, he

---

[116] Dicke (1962b).

[117] Dicke (1961b), p. 100.

[118] On Dicke's life and work, see Peebles (2008). Interviews with Dicke include Lightman and Brawer (1990), pp. 201–213, and three interviews conducted by the American Institute of Physics between 1975 and 1988. The interviews can be found online as http://www.aip.org/history/ohilist/transcripts.html. Dicke reprinted most of his early papers on relativity, cosmology and geophysics in a book-length chapter on "Experimental Relativity" in DeWitt and DeWitt (1964), pp. 165–316. For a review of Dicke's work in geophysics, see Kragh (2015c).

[119] See Kaiser (1998) for aspects of the renaissance.

studied them on his own. In an interview with astrophysicist Martin Harwit he recalled:

> I got interested in astrophysics and geophysics because I thought these subjects provided a tool for getting at some of the questions which my interest in general relativity were bringing up. I was interested early in a number of aspects of relativity, and one of these came about from Dirac's arguments about the large numbers. Another was the interest I had in Mach's principle, thinking this should be significant. ... In that framework you have the requirement that the gravitational constant is not a real constant but it's a function of coordinates. The cosmological solution varies with time, which carries with it obvious geophysical and astrophysical implications, if true.[120]

In 1957 Dicke was appointed professor of physics as Princeton University. The same year he participated in one of the important events of the renaissance of general relativity, the 1957 Chapel Hill Conference on the Role of Gravitation in Physics which on the initiative of Bryce DeWitt and his physicist wife Cécile DeWitt-Morette was held at the University of North Carolina.[121] Other participants included John Wheeler, Peter G. Bergmann, Hermann Bondi, Thomas Gold, Felix Pirani, Stanley Deser, and Richard Feynman. The focus of the conference was on aspects of general relativity, including gravitational waves and problems of quantum gravity, but there was also a session on cosmology. In a summary account of the conference Bergmann briefly referred to "the theories sponsored for instance by P. Jordan," but only to comment that they "have [not] as yet been clarified conceptually to the extent that they merit experimental testing."[122] Jordan's name did not appear in the report of the conference.

Another major conference on general relativity, celebrating the fiftieth anniversary of Einstein's relativity theory, was held in Bern in 1955.[123] Jordan was invited to this event, where he gave a lecture on mathematical aspects of his gravitation theory with $G(t)$. Several of the participants in Bern would also come to Chapel Hill 2 years later, including Bergmann, Deser, Bondi and Pirani. Whereas the Bern conference had been attended by only nine U.S.-based physicists out of a total of

---

[120] Interview of 18 June 1985. See http://www.aip.org/history/ohilist/4572.html

[121] For the background of the Chapel Hill conference 18–23 January 1957, see DeWitt and Rickles (2011), which includes the reports and discussions of the conference. There were interesting links between the Chapel Hill conference and the Gravity Research Foundation (GRF) to be mentioned in Sect. 3.1. Bryce DeWitt won the first prize in the 1953 competition and was instrumental in lifting the scientific respectability of the GRF.

[122] Bergmann (1957). On the other hand, another of Einstein's former German-American collaborators, Valentine Bargmann at Princeton University, covered in the same volume of *Reviews of Modern Physics* Jordan's theory in a review article on the theory of relativity. Bargmann evidently found the possibility that G might vary in space and time to be interesting and worthwhile to test experimentally. See Bargmann (1957).

[123] Mercier and Kervaire (1956). Yet another important event in the renaissance of general relativity (including cosmology) was the 1958 Solvay Congress on "Astrophysics, Gravitation, and the Structure of the Universe," which included addresses by Bondi, Hoyle, Wheeler, Lemaître, and Klein. Two years later the International Committee on General Relativity and Gravitation was established.

ninety or so, Chapel Hill was dominated by the U.S. relativity community. This conference or workshop was smaller and more oriented towards recent work in general relativity than the Swiss conference.

It should be kept in mind that in the 1950s general relativity, cosmology and related aspects of fundamental physics were typically regarded with some disrespect or even hostility by the physics community. Fundamental gravitational physics was seen as something half-way between mathematics and philosophy—not real physics. DeWitt recalled that in the mid-1950s Samuel Goudsmit, the Editor-in-Chief of *Physical Review*, seriously considered banning "papers on gravitation and other fundamental theory" from the journal.[124] In a popular paper of 1961 Dicke said about his own experience:

> As a graduate student of physics 20 years ago I had been told by my professor, a well-known and outstanding physicist, that I should not trouble to learn General Relativity, Einstein's theory of gravitation. As he put it, gravitation was too weak an interaction to be important inside the atom, the site of the big mysteries. This attitude is still mirrored in our graduate training programme, for few universities have even a single graduate course on General Relativity.[125]

Admitting that "the chief support for General Relativity is the simplicity, elegance and beauty of the formalism, rather than observations," Dicke described experimental general relativity as "a sadly neglected field." He was determined to do something about it.

Dicke's talk at Chapel Hill on "The Experimental Basis of Einstein's Theory" mostly dealt with the consequences of a possible decrease in time of the gravitational constant. But he also found time to comment on the recent discovery of violation of parity conservation in beta decay, speculating that "it could conceivably indicate some interaction with the universe as a whole."[126] As to Dirac's "famous dimensionless numbers," Dicke wondered about their meanings. He listed three possible answers:

> First, and what ninety percent of physicists probably believe, is that it is all accidental; approximations have been made anyway, irregularities smoothed out, and there is really nothing to explain; nature is capricious. Second, we have Eddington's view, which I may describe by saying that if we make the mathematics complicated enough, we can expect to make things fit. Third, there is the view of Dirac and others that this pattern indicates some connections not understood as yet. On this view, there is really only *one* "accidental" number, namely, the age of the universe; all the others derive from it. The last of these appeals to me; but we see immediately that this explanation gets into trouble with relativity

---

[124] DeWitt (2009), p. 414. According to DeWitt, Wheeler succeeded in persuading Goudsmit not to keep general relativity out of *Physical Review*.

[125] Dicke (1961c), p. 797.

[126] Dicke (2011), p. 58. The possibility of parity non-conservation in weak interactions had been inferred on theoretical grounds by Tsung Dao Lee and Chen Ning Yang in 1956 and was verified experimentally in early 1957. The discovery was announced at a meeting of the American Physical Society in January 1957, just in time for the physicists at the Chapel Hill conference to know about it.

theory, because it would imply that the gravitational coupling constant varies with time. Hence it might also well vary with position.[127]

As far as the astronomical and geophysical effects of $G(t)$ were concerned, Dicke briefly dealt with the climate in past geological ages, formation of the Moon, and heat flow out of the Earth. A more formal and much more elaborated version appeared half a year later in two papers in *Reviews of Modern Physics*.[128] In one of the papers Dicke proposed a rather speculative theory of gravitation based on an analogy to a polarizable vacuum, which he described by a time-varying vacuum permittivity $\varepsilon_0 = \varepsilon_0(t)$.[129] He suggested that his new picture of empty space—or ether, as he later said—might lead to the creation of particles in an originally matter-free primordial universe containing only gravitational energy. Two years later he spoke about empty space as an ethereal medium: "One suspects that, with empty space having so many properties, all that had been accomplished in destroying the ether was a semantic trick. The ether had been renamed the vacuum."[130]

The other and more interesting of the papers was published under the rather misleading title "Principle of Equivalence and the Weak Interactions." Not only was most of the paper concerned with geological consequences of a varying gravitational constant, it also had nothing to do with weak interactions as usually understood. Dicke took the term to mean interactions described by a small coupling constant, which primarily meant gravitation. His only reference to what is normally called weak interactions was at the end of his paper, where he inferred from Dirac's LNH that "β decay rates would vary inversely as the square root of the age of the Universe." This was the same result that Jordan had suggested in 1944 and also referred to in his 1952 monograph, but Dicke was apparently unaware of Jordan's ideas. A few years later Dicke considered the possible variation in time of the weak coupling constant in relation to the age of meteorites. He suspected the beta decay rate to vary as t^{-n}, with $\frac{1}{4} < n < \frac{1}{2}$, but admitted that available evidence was too uncertain to show any variation.[131]

Like Jordan in Germany, Dicke was fascinated by Dirac's LNH, but in a more critical and independent way. From Dicke's point of view, the numbers of orders

---

[127] Dicke (2011), p. 53.

[128] Issue no. 3 of the 1957 volume of *Reviews of Modern Physics* contained papers prepared in connection with the Chapel Hill conference. Most of the papers, including Dicke (1957b), had not been presented at the conference but may have been informally discussed. The other paper by Dicke (1957a) followed his presentation at Chapel Hill. The issue of *Reviews of Modern Physics* also included Hugh Everett's doctoral dissertation on the "relative state" formulation of quantum mechanics which in Bryce DeWitt's later formulation became known as the many-worlds interpretation. Although Everett did not attend the Chapel Hill conference, his ideas were mentioned in the discussions.

[129] Dicke (1957b). On this topic, see also Sect. 4.1, footnote 1.

[130] Dicke (1959a), p. 29.

[131] Dicke (1959b). See also Peebles and Dicke (1962b). Solomon's suggestion of 1938 resulted in a decay constant varying as t^{-n} with $n = 2/3$ (see Sect. 2.2).

10^{39} and 10^{78} could not have been much different, since they were conditioned by the presence of intelligent life.[132] Humans could not have evolved had the age of the universe T been much smaller than 10^{39} atu, nor would they exist if the age was much greater. Over the next several years he amplified his argument that Dirac's reasoning contained a "logical loophole" by assuming the epoch of humans to be random. "With the assumption of an evolutionary universe, T is not permitted to take one of an enormous range of values, but is somehow limited by the biological requirements to be met during the epoch of man," he wrote.[133] His paper in *Nature* caused a brief reply from Dirac, who now, for the first time since 1938, returned to cosmology.

Dicke's biologically oriented interpretation of the LNH eventually became an important stimulus to the anthropic principle introduced by the Australian-British astrophysicist Brandon Carter in 1973.[134] In his widely read textbook *Cosmology* from 1952, Hermann Bondi included sections on Dirac's large numbers, about which he said that "it is difficult to resist the conclusion that they represent the expression of a deep relation between the cosmos and microphysics."[135] Although Bondi thus was receptive to the magic of the LNH, he concluded that Dirac's cosmology based on it was "very unconvincing" and that Jordan's theory was "almost certainly false." It was by reading Bondi's book that Carter became interested in the line of reasoning that led him to the anthropic principle. By 1970 Carter had become acquainted with Dicke's work and Dirac's LNH but not yet formulated the anthropic principle. He rejected Dirac's $G(t)$ conclusion as "an error of blatant wishful thinking," as he later expressed it, and informally he discussed what he at the time called the "principle of cognizability."[136]

The question of the Earth's temperature in the past first raised by Teller was reconsidered by Dicke, who dealt with it in greater detail than previous authors. Assuming the age of the universe to be 8 billion years, in 1962 he calculated the surface temperature of the Earth during the last 4 billion years.[137] Rather than using Dirac's $G(t)$ variation he adopted the slower variation as given by the Brans–Dicke theory with $\omega = 6$, namely

---

[132] Dicke (1957a, p. 356, 1957b, p. 375).

[133] Dicke (1961a, 1959a).

[134] Barrow and Tipler (1986). On the early history of the anthropic principle, see Kragh (2011), pp. 220–228.

[135] Bondi (1952), pp. 59–62, 159–164.

[136] Carter (1989), p. 190. Dyson (1972) mentioned Carter's "principle of cognizability" in relation to Dirac's $G(t)$ hypothesis. The anthropic principle was introduced in Carter (1973), where Carter stated it as the opposite of the Dirac–Jordan varying-G hypothesis.

[137] Dicke (1962a). See Sect. 3.6 for a version of Dicke's temperature-time graph.

$$\left(\frac{1}{G}\frac{dG}{dt}\right)_0 = -\frac{H_0}{7} \cong -1.2 \times 10^{-11}\,\text{year}^{-1}.$$

Moreover, Dicke took into account the great amount of water vapour in the cloud-covered atmosphere caused by increased temperature. The effect of the water vapour, he suggested, would be to stabilize the temperature. As a result of his calculations he suggested that the surface temperature in the early period of the history of the Earth agreed with the existence of algae some 3 billion years ago. Such algae were known to survive in hot springs at a temperature of nearly 90 °C. Dicke's analysis confirmed his earlier conclusion that "there is no particular difficulty in accounting for life over a period of the past billion years."[138] The conclusion presupposed a variation of G of the Brans–Dicke type and not of the Dirac–Jordan type $G \sim 1/t$, where the rate of decrease of G would be considerably larger, namely 2.4×10^{-10} year^{-1}.

In a review of 1961 Dicke estimated the past surface temperature of both the Earth and the Moon on the assumption of Dirac's $G(t)$ and a Hubble constant of $H_0 = 80$ km s^{-1} Mpc^{-1}. Although he admitted that the curve for the Earth's variation with temperature "has no great reliability," he was confident that the highest temperature for the Moon would have been some 250 °C at the time of its formation.[139]

By the late 1970s climate models indicated that the solar constant had been remarkably constant over a period of 3×10^9 years and that the change in solar luminosity during the last 300 million years had been less than 3 %. Most experts now agreed that the brightness of the Sun was of relatively little importance for variations in the past climate of the Earth.[140] It appeared that questions of paleo-climatology were too messy and complex to be answered by the traditional methods of physicists and astronomers. Nor was paleoclimatology of any real use in testing competing cosmological models. The conclusion of Dicke was agnostic and some-what despairing: "The moral is that the atmosphere is complicated . . . We cannot be sure how much the surface temperature would have changed."[141] We shall return to the question of the ancient Earth's temperature in Sect. 3.6.

---

[138] Dicke (1957a), p. 358.

[139] Dicke (1961b, p. 101, 1964b, p. 160).

[140] Wigley (1981).

[141] Dicke (1964b), p. 160.

Chapter 3
The Expanding Earth

In the 1960s the once discarded theory of continental drift proposed by Alfred Lothar Wegener was substantially revised and transformed into the modern standard theory of global plate tectonics. For a decade or so the new theory of the Earth and the traditional contraction theory faced competition from a third alternative, the hypothesis of the expanding Earth. As early as 1952 Jordan had suggested Earth expansion on the basis of decreasing gravity, and a few years later the suggestion was taken up by several physicists and earth scientists. Dicke seriously applied his skills in fundamental physics to a broad range of geophysical problems, including a possible increase in the Earth's radius. The Hungarian geophysicist László Egyed was not only a leading figure in the expansionist alternative but also an advocate of varying gravity as the cause of the growing Earth. Other geologists and geophysicists in favour of the expanding Earth preferred to present their chosen theory in purely empirical terms, without considering the cause of the expansion.

This chapter offers a fairly comprehensive history of the expanding Earth hypothesis in the period up to the 1970s with an emphasis on those scientists who considered expansion to be connected with Dirac's hypothesis of a decrease in the gravitational constant. The later phase of the two hypotheses will be examined in Chap. 4.

3.1 Drifting Continents and the Expansion Alternative

Alfred Wegener's epoch-making theory of continental drift and its fate in the decades following its conception in 1912 is thoroughly covered in the historical literature.[1] Some of the ideas were anticipated by the American geologist Frank

---

[1] The literature includes Hallam (1973), Menard (1986), Le Grand (1988), Oreskes (1999), and Frankel (2012a).

© Springer International Publishing Switzerland 2016
H. Kragh, *Varying Gravity*, Science Networks. Historical Studies 54,
DOI 10.1007/978-3-319-24379-5_3

B. Taylor a few years earlier and also the Irish geologist John Joly has been mentioned as a predecessor. Yet it was only with Wegener—and especially with his classic monograph *Die Entstehung der Kontinente und Ozeane* (The Origin of Continents and Oceans) from 1915—that a coherent, logical and empirically argued theory of drift was presented as an alternative to the established "fixist" view of the Earth. According to this traditional view, the Earth was cooling and essentially static in spite of a slow thermal contraction. Wegener, on the other hand, assumed that in the Cretaceous period the original super-continent Pangaea divided into two continents, and that subsequent divisions and motions resulted in the globe as we know it today. The continents not only separated, they literally ploughed through the ocean floor.

Wegener's theory was much discussed in the 1920s when it was seen at the same time as revolutionary and controversial. For a brief period of time it was seen as kindled in spirit by other fashionable theories and trends, such as Einstein's relativity, Picasso's cubism, and Freud's psychoanalysis. Mobilism was *à la mode*, but not for long.[2] While Wegener's ideas aroused enthusiasm in some quarters, the general reaction among geologists and geophysicists was marked by scepticism or downright hostility. Apart from empirical weaknesses, a common reason for dismissing the hypothesis of continental drift was that Wegener was unable to provide the postulated motion of the continents with an adequate physical cause. Wegener admitted the phenomenological nature of the hypothesis. As he wrote, alluding to planetary astronomy at the time of Kepler, "The Newton of drift theory has not yet appeared."[3] The problem was not only that there was no known cause but that drift seemed to be impossible as it contradicted the laws of physics. To the British mathematician and geophysicist Harold Jeffreys, a formidable and influential opponent of continental drift, this was a decisive argument against Wegener's theory.

Despite a generally cool or hostile response, Wegener's theory was not without supporters. The South African geologist Alexander du Toit adopted, developed and promoted drift theory during the 1930s and 1940s when it was abandoned by most of his colleagues. Even more important was Arthur Holmes, who in the late 1920s converted to continental drift and proposed a plausible mechanism of the motion of the continents based on radioactive heating and convection currents in the Earth's mantle. In part inspired by the earlier work of Joly, in 1925 Holmes reached the conclusion that "it is no longer possible to resist Wegener's intriguing displacement theory on the grounds that the continental blocks are at the present day embedded in a rigid substratum."[4]

All the same, the support of du Toit, Holmes and a few others did not succeed in turning drift theory into an appealing alternative that enjoyed broad recognition in the community of earth scientists. That happened only in the late 1950s when a

[2] See Wood (1985), pp. 71–76.
[3] Wegener (1966), p. 167.
[4] Holmes (1925), p. 531.

growing number of geophysicists and marine geologists began to realize the advantages of continental drift and similar mobilist ideas. A major reason for the changed fate of drift was advances in paleomagnetism and the understanding of the ocean floors. The resurrection, transformation and extension of Wegener's theory that led to global plate tectonics in the 1960s were to a large extent due to geophysicists, marine geologists and oceanographers. Classical geologists, many of whom were unfamiliar with paleomagnetism and other parts of geophysics, were not in the vanguard of the revolution.

When Jordan proposed the idea of an expanding Earth in 1952, he believed it was an original idea. Only a few years later did he realize that the general idea of an Earth that expands rather than contracts over time can be found much earlier.[5] The first proponent of Earth expansion may have been an Englishman, William Lowthian Green, who introduced the idea in 1857. During the next 80 years or so several authors, especially in Russia and Germany, came up with suggestions of an expanding Earth. These early theories were typically of a speculative and amateurish nature, and they mostly appeared in obscure publications unknown to professional geologists. They made almost no impact at all on post-World War II development. Halm's astronomically based theory of 1935, the essence of which is outlined in Sect. 1.3, was an exception in the sense that it was published by a respected astronomer in a recognized if not widely circulated scientific journal. And yet Halm's theory made no more impact than those of the more speculative authors.

The only one of Jordan's predecessors deserving mention is the German engineer and amateur geophysicist Ott Hilgenberg who in 1933 self-published a book with the title *Vom Wachsenden Erdball* (On the Growing Earth). It was dedicated to Wegener. However, although convinced of Wegener's arguments for continental drift, Hilgenberg did not accept his picture of the horizontally moving continents. As an alternative he argued that the continents were driven apart from each other as a result of a radial expansion of the Earth. He thought that the Earth had expanded rapidly since the early Mesozoic. Hilgenberg's theory shared a feature of several other pre-World War II speculations, namely that the Earth not only grew in size but also in mass. His idea was completely out of tune with physics at the time as it rested on the belief that ethereal energy from the cosmos was continually absorbed in "ether sinks" and from there transformed into matter.[6] When Hilgenberg cannot be dismissed as just a crank, it is for two reasons. First, many of his ideas anticipated the later scientific expansion models. Second, in the 1960s he returned to Earth expansionism and after his death in 1976 he even achieved status as an

---

[5] Jordan (1955, p. 226, 1971, p. 72). On early Earth expansionism, see Carey (1988), pp. 137–141, and Nunan (1998).

[6] Hilgenberg may have borrowed the idea of ether sinks from the British mathematician and philosopher Karl Pearson, who in the 1890s developed a theory of matter and ether based on "squirts" and "sinks." See Kragh (2011), pp. 39–40.

ancestor hero among some expansionists.[7] His obscure and uninfluential book of 1933 was now considered a classic and visionary work of the earth sciences.

From the mid-1950s onwards the idea of an expanding Earth was discussed and sometimes advocated by several earth scientists either as a supplement to or, more commonly, a substitute for the hypothesis of drifting continents. It is worth emphasizing the main difference between the two hypotheses. According to continental drift there had always been continents and oceans, but their patterns of distribution have changed as the continents separated on the surface of the constant-sized Earth. By contrast, there were no oceans in the expansionists' picture of the original Earth, which was completely covered by a sialic crust. Only with expansion of the Earth and the resulting cracks in the crust did the oceans appear. Expansionists disagreed about the finer details of the history of the Earth, including the rate of the expansion and its beginning in geological time, but all agreed that the continents had separated as a result of an increased size of the globe (Fig. 3.1). The pioneers of the new expansionism were Jordan from Germany, Egyed from Hungary, Carey from Australia, and Heezen from the United States.

For a decade or so "expansionism" appeared as a possible and even, in some quarters, attractive alternative to both the traditional contraction theory and the emerging plate tectonics based on continental drift.[8] However, by the early 1970s expansion was considered a dead issue by the majority of geologists and geophysicists. The absence of a convincing mechanism for the slow inflation of the Earth was an important reason why, in the end, expansion theory was broadly dismissed. Recall that Wegener's inability to provide a satisfactory mechanism for drifting continents was a main reason for dismissing the old drift theory.

Most supporters of an expanding Earth relied on empirical arguments alone without caring much for the cause of the expansion. "I personally have no strong feelings concerning the specific mechanism of expansion," said Bruce Heezen in 1960, "I simply conclude from the morphological and paleomagnetic results that expansion has occurred."[9] The prominent expansionist Warren Carey likewise stated that "Empirically I am satisfied that the earth is expanding," adding that he

---

[7] Scalera and Braun (2003).

[8] The expansion theory of the Earth is discussed in relation to plate tectonics in, for example, Le Grand (1988), pp. 193–195, Menard (1986), pp. 142–151, and Oldroyd (1996), pp. 273–278. From a more philosophical than historical perspective the theory is dealt with in Nunan (1988), whereas Nunan (1998) provides a concise summary of the history of the hypothesis. Sudiro (2014) focuses on expansion theory's degeneration in recent time into what he argues, probably correctly, is a pseudoscience. Holmes (1965) includes a useful semi-historical chapter (pp. 960–994) on the hypothesis. Carey (1976, pp. 23–38, 1988) are informative if partisan accounts of the development of earth expansionism. See also the (no less partisan) bibliography in Scalera and Jacob (2003), pp. 419–421. The most detailed and scholarly work on the subject is contained in Henry Frankel's four-volume work on the history of plate tectonics, see Frankel (2012b), especially pp. 278–354. Other histories of plate tectonics tend to ignore the expanding Earth. For example, the theory is not mentioned in Oreskes (2001), a collection of essays written by participants in the plate tectonics revolution.

[9] Quoted in Frankel (2012c), p. 417.

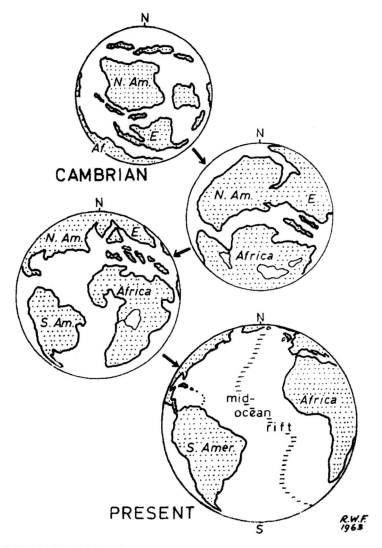

Fig. 3.1 Rhodes Fairbridge's paleogeographic sketch of the expanding Earth. *Source*: Fairbridge (1966), p. 144. Reproduced with the permission of John Wiley and Sons

found none of the proposed explanations to be satisfactory.[10] But other scientists did suggest mechanisms, either of a terrestrial or a cosmological nature. The hypothesis of a decreasing gravitational constant, as advocated principally by Jordan, Egyed and Dicke, obviously belonged to the latter category, which is the

---

[10] Carey (1975), p. 134.

one of most interest in the present context. The $G(t)$ mechanism was well known to geophysicists in the 1960s but far from generally accepted.

Samuel Warren Carey, professor of geology at the University of Tasmania, had defended continental drift since the 1930s and as early as 1953 he considered subduction of the oceanic crust. According to the leading geologist Edward Irving, "It is possible that if he [Carey] had held to a fixed radius he would have formulated the main ideas of plate tectonics before 1960—before any one else."[11] However, a few years later Carey switched to expansionism. At a symposium held in March 1956 in Hobart, the capital of Tasmania, he reported that it was only possible to assemble Pangaea, Wegener's original supercontinent, on an Earth considerably smaller than its present size.[12] The new picture of the Earth that Carey outlined was this:

> At an early stage of the earth's history the crust was uniform. . . . The diameter of the earth was less than half its present diameter and its surface area less than a quarter of its present surface. The mean density was more than eight times the present density or over 44 [g cm^{-3}]! Surface gravity was four times the present gravity. The rate of rotation was correspondingly great.[13]

The hypothetical ocean called Tethys was part of the drifting continents tradition, but not in Wegener's version which postulated a single landmass (Pangaea) as a starting point. Alexander du Toit assumed in his version two supercontinents, the southern Gondwana and the northern Laurasia which were separated by Tethys. He took over the names and the idea from the great Austrian geologist Eduard Suess, according to whom Thetys was an inland sea. Whatever the origin of the concept, according to Carey and most other expansionists, the original Tethys ocean was just an artefact. It had never existed, for the Earth started its expansion without any oceans at all. The oceans only came into existence as the Earth grew bigger.

As Carey saw it, the Earth's radius R had increased since the Paleozoic at an average rate of 0.5 cm per year, but at the end of the Carboniferous, when R was about 5000 km, the expansion had begun accelerating. Carey eventually reached the conclusion that at the Ordovician $R \cong 0.6\,R_0$ and at the late Paleozoic $R \cong 0.7\,R_0$, where R_0 refers to the present radius.[14] Recognizing that it was difficult and perhaps impossible to explain the rapid expansion by means of the known laws of physics, he stressed that it was an empirical conclusion. All the same, he speculated that it might be due to a new kind of phase change in the interior of the Earth, where high pressure forced the inner shells of electrons to merge with the atomic nucleus. This kind of explanation sketch had been foreshadowed by Halm in his paper of 1935.

---

[11] Irving, in a review of Carey (1976) appearing in *Tectonophysics* **45** (1978): 241–242.

[12] Carey (1958) discussed his expansion hypothesis in the proceedings of the Hobart symposium, which was however only published two years after the symposium itself. There are reasons to believe that he did not discuss the issue in his talk and that his conversion thus took place a little later. See Frankel (2012c), pp. 322–335.

[13] Carey (1958), p. 346.

[14] Carey (1983), p. 181.

If the cause of the expansion could not be explained by the laws of physics, Carey believed it would be a problem for physicists rather than geologists. It might even be "the clue leading to new fundamental developments in physics," he suggested.[15]

Although Carey may at the time have been aware of Dirac's $G(t)$ hypothesis, neither he nor others mentioned it at the Hobart symposium. Carey would do so in his later publications, but without ever embracing it. Over the following decades Carey emerged as a charismatic and controversial leader of rapid global expansion, proselytizing the idea at every possible occasion. At the same time he increasingly estranged himself from mainstream geophysics.

3.2 Pascual Jordan: Geophysicist?

Apart from references to the age of the Earth, in his work until the 1950s Jordan did not relate Dirac's hypothesis to issues of geology or geophysics. However, from that time onwards he increasingly focused on the Earth as a testing ground for cosmological theories in general and for the $G(t)$ hypothesis in particular. At a conference on cosmological models held in Lisbon in 1963, he said:

> Our Earth is the only [celestial body] about which we can perform any *detailed historical research*; and the earth will maintain this singular importance in the frame of our knowledge as long as space travel to the moon and to other planets did not yet become part of the everyday life. The history of other celestial bodies is a matter of theory (and often only of speculation). Only the earth allows us detailed sure statements. Therefore there cannot be seen anything surprising, but only a very natural way of approach if we consider as *necessary* sources of cosmological information also such branches as geophysics, geology, paleomagnetism, paleoclimatology.[16]

Contrary to the second edition of *Schwerkraft*, which appeared in 1955, the original edition was little known and rarely referred to outside Germany. The reason for the early lack of interest in Jordan's work may in part have been due to external factors, such as the decline of German as an international language of physics and a partial interruption of scientific communication between Germany and the Western countries in the years following World War II.[17] Jordan's theory was not entirely unknown, though. Significantly, Dicke referred to both editions of *Schwerkraft*. Moreover, the book of 1952 was reviewed by two of the leading British cosmologists, namely by George McVittie in the *Proceedings of the Physical Society* and by William McCrea in the highly visible *Nature*. In both cases the

---

[15] Carey (1958), p. 349.

[16] Jordan (1964), p. 111. See also Jordan (1971), pp. 9–10, where he discussed the question, "Why should the Earth, a single planet among the many millions of celestial bodies in the Milky Way, be a decisive test object for a fundamental physical law?"

[17] Goenner (2012). Jordan's Nazi past may also have played a role.

reviews were rather critical and they did not relate at all to Jordan's idea of an expanding Earth.[18]

The second edition of *Schwerkraft* differed from the first one in several respects. It was revised in collaboration with Engelbert Schücking (or Schucking), who earned his Ph.D. degree in Hamburg in 1955. Schücking collaborated with Jordan and his group on aspects of general relativity until 1961, when he moved to the United States to become professor of physics at New York University six years later.[19] Perhaps the most important difference was that the second edition of *Schwerkraft* included a more detailed, if largely qualitative treatment of the geophysical consequences of the varying-*G* hypothesis. A new section on these consequences dealt with problems that, in Jordan's opinion, ought to "attract attention among astronomers, geophysicists, geologists, and paleoclimatologists." However, Jordan was keenly aware that in the area of the earth sciences he was an amateur: "The author deals with subjects that lie far outside his professional competence and for this reason his accounts are highly defective."[20] He realized that his geophysical evidence for a decreasing *G* was of an indirect nature only and that it might not appear convincing to the geophysicists. He consequently stressed that it was still somewhat uncertain if the geological and astronomical issues he referred to were really evidence in support of his theory of a decreasing *G*.

It was indeed unusual for a mathematically inclined quantum physicist to take up problems of geology, but it was not the first time that Jordan made the move from theoretical physics to areas of natural history. "I have always regarded the true aim of my life as lying in the activity of a natural scientist rather than that of a pure physicist," he stated.[21] While a student at the University of Göttingen he studied not only physics and mathematics, but also zoology. And in the 1930s he engaged seriously in borderline problems between quantum physics and biology, attempting to create a theory of so-called quantum biology.[22] Now Jordan invested much of his intellectual resources in the earth sciences, well knowing that he did it as a layman. But this was not necessarily a disadvantage, he thought:

---

[18] Dicke (1957a, b). McVittie's review appeared in *Proceedings of the Physical Society* (London) A **66** (1953): 667–668, and McCrea's in *Nature* **172** (1953): 3–4. Strangely, McCrea wrote about the *G(t)* hypothesis that it "was once briefly suggested and later discarded by Dirac." He probably meant the *N(t)* hypothesis of spontaneous creation of matter. An extensive English review of Jordan's theory appeared in Brill (1962). Noting that Jordan's theory "has so far not received any great attention," Harrison (1963) referred to Jordan (1949, 1952).

[19] See Schucking (1999) and Ehlers and Schücking (2002). These sources give some background on Jordan and the roots of his theories of gravitation and the expanding Earth.

[20] Jordan (1955), p. vi and p. 223.

[21] Jordan (1971), p. x.

[22] Beyler (1996). On Jordan's philosophy of science and general world view, see Jordan (1963), a popular book which includes sections on varying gravity and the expanding Earth. See also Kragh (2004), pp. 175–185 and Beyler (2009). Beyler (1994) deals with Jordan's biological and cosmological work in the post-World War II period, but does not mention his extensive work on the expanding Earth and other aspects of geophysics.

I believe it to be a healthy activity for a research scientist to take up a supplementary activity besides the great amount of specialist work. He should aim at studying the various points of contact of different special fields, and give much thought to the co-ordination of different research activities and their results. Very often the specialist is not sufficiently protected against the suggestive power of ad hoc hypotheses, whose weaknesses can be seen much more easily and clearly if viewed from other subjects.[23]

When comparing the earth sciences to physics, Jordan was unimpressed. His extensive but not very systematic reading of the geological, geophysical and oceanographic literature left him with "the impression of a multitude of contradictory theories" that caused more confusion than enlightenment. The physicist, he wrote, "is disconcerted to find that the ratio of facts to hypotheses in, for example, geology is very different from that in physics."[24] At a later occasion, a symposium in honour of Dirac's seventieth year birthday, he ascribed his own approach to what he called the different mentalities of physicists and geologists, suggesting that the latter favoured a conservative style of thinking. Contrary to the physicists they were not eager to learn new facts, he claimed. With this in mind,

I came to the decision not to ask other specialists [in the earth sciences] about the compatibility of Dirac's hypothesis with empirical facts, but to try to learn myself what really are the proven facts, and to see whether they lead to real contradictions against Dirac's hypothesis.[25]

Prepared to find no contradictions, Jordan found none.

In a letter of 1 October 1952 the distinguished theoretical physicist Wolfgang Pauli, a Nobel laureate of 1945 and one of Jordan's old colleagues and collaborators, thanked him for a copy of *Schwerkraft und Weltall*. Pauli had little interest in cosmology and none in geophysics, but he was an authority in the area of relativity theory and did not consider the $G(t)$ hypothesis to be particularly heretic. "In itself Dirac's idea of a varying κ [$=8\pi G/c^2$] seems natural to me," Pauli wrote, "but as yet I dare not form an opinion of whether or not it corresponds to physical reality."[26] In a later letter from the same year Jordan referred to his new interest in geophysical evidence for a gravitational constant decreasing in time. After having told Pauli about his recent work on the extended scalar–tensor theory of general relativity, he said:

---

[23] Jordan (1971), p. xi. Referring to geologists and geophysicists, Jordan claimed on p. 141 that "[it is] the practice of many authors first to put forward a definite theory ... and then to discuss the empirical facts in terms of this theory." This procedure he much disliked because it "cannot produce conclusions derived unambiguously and logically from the existing foundations."

[24] Jordan (1971), p. 10. See also Jordan et al. (1964), p. 506. The eminent geophysicist J. Tuzo Wilson noted the different research styles of physicists and geologists in their studies of the Earth, but he viewed them in a different light than Jordan. "Physicists' generalizations have tended to be too sweeping and geologists' too detailed," he wrote. Wilson (1963b), p. 864.

[25] Jordan (1973), p. 61. On Jordan's lack of respect for the geological literature, see also Jordan (1969b), p. 260.

[26] Pauli (1996), p. 736.

I amuse myself with noting that various empirical facts indicate that the gravitational constant was larger a few billion years ago than it is now. In my book I have referred to the ideas of the American Fisher that the composition of the surface of the Earth seems to indicate that the surface has increased by a factor 2 to 3 since the Earth was formed. One would expect a similar effect on the Moon, although by a *much* smaller factor. In fact, in the youngest of its formations the Moon exhibits certain "rills" which until now have remained unexplained ... and [which] give the strong impression that the interior of the Moon subsequently has experienced a very small expansion in volume. There are some more facts that in my view deserve a closer discussion. But an attempt to deal with this question by means of quantitative calculations can be made only by specialists in geophysics such as Bullen; and they will find my idea too incredible to care about it.[27]

Keith Edward Bullen was a New Zealand-born geophysicist, seismologist and applied mathematician who in the late 1940s inferred from seismological data that the inner core of the Earth is solid and with a higher density than the outer core. He was one among several geophysicists who established the modern view of the interior of the Earth.[28] Jordan was familiar with Bullen's so-called "compressibility-pressure hypothesis" and his calculations of the density of the Earth's inner core.[29]

A leading theme in the 1955 edition of *Schwerkraft* was the hypothesis of an expanding Earth as a result of the decreasing gravitational force. It seems that Jordan came to the hypothesis from an unlikely source, a wealthy American businessman and amateur scientist by the name Joel E. Fisher.[30] Mr. Fisher first appeared in Jordan's publications in the first edition of *Schwerkraft* and subsequently in all of his work related to the expanding Earth. The two probably met personally, but the circumstances are unknown. Their first interaction may have taken place in the early 1950s in the form of correspondence in relation to Jordan's forthcoming book on gravitation and cosmology.[31]

Joel Ellis Fisher, born 30 November 1891, was president of the North-Western Terminal Railroad in Denver and later director of the Melville Shoe Corporation in Harrison, New York.[32] His active business career made him a wealthy man who privately sponsored research on various subjects, including glaciers and gravitation. Before he entered the world of private business Fisher had studied at Yale from where he graduated in 1911. He was director of the Babson Gravity Research Foundation and the Washington Institute for Biophysical Research. Since his

---

[27] Letter of 17 December 1952, in Pauli (1996), p. 800. For "the American Fisher," see below.

[28] Bullen (1949). On Bullen's work on the core of the Earth, see Brush (1996c), pp. 198–202.

[29] Jordan (1952), p. 198.

[30] Jordan (1955), p. vi and p. 226.

[31] Jordan (1961a), p. 417. Jordan referred in some of his writings, for example Jordan (1961b), to discussions with Fisher.

[32] Obituaries of Fisher appeared in *Alpina Americana Journal* **15** (1966): 115–116 and *Alpine Journal* **71** (1966): 190. These sources can be found online as http://publications.americanalpineclub.org/articles/12196611500/Joel-Ellis-Fisher-1891-1966 and http://www.alpinejournal.org.uk/Contents/Contents_1966_files/AJ%201966%20190-198%20In%20Memoriam.pdf. To my knowledge, the Jordan–Fisher connection has never been noted in either the scientific or historical literature.

youth Fisher was a devoted mountaineer and member of several American and European mountaineering clubs. During the late 1930s he served as president of the Alpina Americana club. He climbed the Matterhorn in Switzerland six times, the last time in 1950, and in 1965 the 74-year-old Fisher ascended the 2930 m high Riffelhorn also in Switzerland. He died in New York on 6 January 1966.

Little is known about the scientific work of Fisher except that he had an interest in orogeny and glaciology, on which subjects he published a couple of papers in the *American Journal of Science* in the 1940s and also, in 1950 and 1955, in the *Journal of Glaciology*. Apart from this he did not publish in recognized journals of physics or geophysics, but privately he published several books, pamphlets and essays on glaciology, mountain ascents, and other geological subjects.[33] Some of these were of a rather speculative and unorthodox nature. One of his privately printed publications was a booklet from 1950 entitled *Some Problems of Geophysics, Approached from Viewpoints of Modern Physics*. In the late 1950s Fisher collaborated with William J. Hooper, a physics-trained inventor, in experiments on the generation of anti-gravity from electric fields. Their research was presented at meetings of the American Physical Society and resulted in two US patents to Hooper.

A serious amateur scientist, in 1949 Fisher was elected a member of the New York Academy of Sciences. Although Jordan referred frequently and very positively to Fisher in his publications from 1955 to about 1970, he only once cited a paper by him.[34] Characteristically, this was a privately printed essay, which only appeared as an abstract in the *Transactions* of the American Geophysical Society. Despite Fisher's lack of reputation in scientific circles, Jordan valued his insights highly and perhaps somewhat uncritically. He never missed an opportunity to express his indebtedness to his friend in New York. As far as geology and geophysics were concerned, it was one amateur being captivated by the ideas of another amateur.

In 1954 Jordan wrote a brief essay on his varying-G theory to the prize competition of the Gravity Research Foundation, a private organization founded 5 years earlier. Gravity Research Foundation (GRF) was established in 1949 by the American economist and businessman Roger W. Babson with financial support from Clarence Birdseye, an inventor and industrialist. The original purpose was to stimulate research in anti-gravity devices, materials that could absorb gravity and other technological applications of gravity. Babson tended to see gravity as "our enemy number one," possibly because he felt that this sinister force of nature was responsible for his sister's and grandson's drowning. His aim with GRF was not so much to study gravitation from a scientific perspective as it was to tame gravity and turn it into a friend rather than enemy of mankind (Fig. 3.2).

---

[33] For some of Fisher's publications and abstracts, see King et al. (1965), p. 537, which includes an abstract titled "Arguments for a solid core of the Earth at 0° K."

[34] See Jordan (1955), p. 223.

Fig. 3.2 Monument to R. W. Babson at Gordon College, Massachusetts. The inscription reminds students of "the blessings forthcoming when science determines what gravity is, how it works, and how it may be controlled." Photograph by Elizabeth B. Thomsen, 2013. Retrieved from https://commons.wikimedia.org

Fisher had close connections to GRF, for which he for a period served as acting director. On 27 August 1960 GRF sponsored a "Gravity Day" with Fisher presenting a paper on "The Possibility of Producing Changes in the Gravitational Mass of Certain Substances." He reported on experiments with bismuth and other elements that apparently proved that they gained or lost weight according to their magnetic history.[35] It took a couple of years until GRF turned to more mainstream research in gravity and then succeeded in attracting interest and respect from the community of physicists and astronomers. In this way GRF became part of the renaissance of general relativity and gravitation studies. Among the prize winners of the 1950s and 1960s were notables such as Thomas Gold, John Wheeler, Banesh Hoffmann, Hermann Bondi, Roger Penrose, Stephen Hawking, and Dennis Sciama. The most recent prize winner (for 2015) is Gerard 't Hooft of Utrecht University, the Netherlands, a Nobel physics laureate of 1999.

The essay to GRF was actually submitted on Jordan's behalf by Fisher, whom Jordan credited for most of his geological examples supporting the theory: "It was first brought to my attention by my friend, Joel E. Fisher, that there exist many well known facts in geology and geophysics completely unexplained, but readily explainable under the proposal that the constant of gravitation has diminished

---

[35] See http://www.presidentialufo.com/wilbert-smith-articles/131-gravity-day-1960. For the history and activities of GRF, see http://www.gravityresearchfoundation.org/ and also DeWitt and Rickles (2011), pp. 7–15.

#3

THE THEORY OF A VARIABLE "CONSTANT" OF GRAVITATION

There exists, then, one Theory of Gravitation originating out of Einstein's own Theory which we shall now describe. This latter, which I have called the "Generalized Theory" (in German, "Erweiterte Gravitationstheorie") does not declare Einstein's theory to be false and inapplicable as to gravitation. But it does declare that a certain degree of generalization of his theory is necessary to explain all phenomena which are related to gravitation. If our Hypothesis is accepted, Einstein's Theory is too restricted - it must be "generalized".

The fundamental idea behind this new theory is the definition of f, the Newtonian Constant of Gravitation, in the Theory of Relativity, by the expression

(1) $$G = \frac{8\pi f}{c^2}$$

(where c is the velocity of light
(3 x 10^{10} cm/sec^2
(and G, the universal "constant" of gravitation
(6.67 x 10^{-8} dyne/cm^2/gm^2)

In agreement with Newton, Einstein admits that G is a true constant, unchanging in value, throughout space, and over all time.

Instead of this, we assert the Hypothesis

G is a variable.

G does vary, with time, and G may also vary from place to place.

Fig. 3.3 Part of Jordan's 1954 essay to the Gravity Research Foundation. Reproduced with the permission of the Gravity Research Foundation

over geological time."[36] Jordan vaguely suggested that his theory, contrary to Einstein's, allowed in principle "the harnessing of gravity," which supposedly resonated with the ideas and aims of the Gravity Research Foundation (Fig. 3.3). However, he did not win the prize but only received honourable mention. He was no more fortunate thirteen years later, when he again submitted an essay, this time on empirical tests of the $G(t)$ hypothesis. Jordan now focused on astronomical measurements rather than on the more uncertain results obtained from geology and geophysics. Relating to his original interest in Dirac's hypothesis, he wrote:

But though mathematically beautifully relations were revealed by these studies … it remained unsatisfactory, that the extreme slowness of the surmised decrease seemed to leave scarcely any hope to connect these mathematical speculations with empirical facts. Therefore I was again fascinated as my late friend J. E. Fisher at New York made the remark, that Dirac's decrease of G, if existent, must have caused a marked expansion of the earth in the course of its history.

---

[36] Jordan (1954).

In works between 1955 and 1962 Jordan discussed a long series of geophysical, geological and astronomical evidence which, in his view, amounted to convincing support of the $G(t)$ hypothesis.[37] In accordance with Fisher he proposed that the Earth must have expanded and that the cause of the expansion was a steadily diminishing gravity. His basic argument was that, since the Earth consisted of a compressed core and a non-compressed crust, diminishing gravity would cause the core to expand as a result of reduced weight of the overlying layer of rock. Stated somewhat differently, with a slow change in G the Earth can be assumed to be in a state of hydrostatic equilibrium at any given time. In such an equilibrium state the pressure p varies with the radius r as

$$\frac{dp}{dr} = -G(t)\frac{\rho m(r)}{r^2}.$$

The quantity $m(r)$ is the mass inside r, and ρ is the density. A decrease in G implies that the pressure in the interior was formerly greater than it is today and, consequently, that the Earth was smaller. According to Jordan, as a result of expansion, the crust would break up and the cracks would be filled in by upward movement of the underlying molten basalt. The total area of granitic continents today would thus be equal to the surface of a smaller ancient Earth. Jordan was pleased to note that the Russian geologist I. B. Kirillov "has shown that the present continents do fit together on a sphere whose surface area is equal to the total area of the continents."[38]

The same point was independently made by Cyril H. Barnett, a British anatomist and amateur geologist. Barnett's argument caused a brief comment from Harold Jeffreys, who noted "the problem of explaining how the Earth's volume can have increased threefold."[39] He further objected that the shape of the continents, had they originally covered the entire Earth, would have been considerably distorted as a result of the expansion. Jeffreys had no more confidence in the expanding Earth than he had in continental drift.

At a conference in Newcastle in 1967 Jordan once again referred to Fisher as the originator of the connection between the $G(t)$ hypothesis and the expanding Earth. As he recalled, having digested Fisher's idea he believed to see in it a possible answer to "one of the great problems of Earth research," namely, why there exists two preferred levels of elevation on the surface of the Earth separated by about 4.5 km. It was then that he "began to read a little about the sciences concerning the Earth and the Moon."[40] Jordan similarly highlighted the idea of a two-layered Earth

---

[37] Jordan (1955, 1959a, b, 1962a, b, c).

[38] Jordan (1962a, b, c), p. 599. See also Jordan (1971), pp. 72–75.

[39] Barnett (1962) followed by Jeffrey's untitled comment. Jeffrey's objection of continental distortions was countered by Dennis (1962), an American geologist. See also Barnett (1969).

[40] Jordan (1969a), p. 55. On the hypsographical problem as a motivation for Jordan's ideas of the expanding Earth, see also Jordan (1969b), pp. 262–263.

in a contribution to a volume published on the occasion of the sixtieth birthday of
the Polish physicist Leopold Infeld. We have to assume, he said, that

> ... during the formation of the continental layer the Earth was considerably smaller than
> today, so that this layer covered the whole sphere of the Earth to a practically constant
> thickness. A process of expansion caused the continental layer to split into parts; the newly
> formed rifts were filled out to the equilibrium altitude by the more dense but also more
> mobile sima material of the deeper layer. In the course of time part of these rifts between
> the parts of the continental layer developed into wide oceans.[41]

The "hypsographical" problem relating to the two-layer structure of the Earth's
surface, corresponding to the continental plates and the ocean floors, was far from
new. For example, it had been dealt with in detail by Alfred Wegener, who believed
it could be explained by continental drift. "In the whole of geophysics," Wegener
wrote,

> ... there is probably hardly another law of such clarity and reliability as this—that there are
> two preferential levels for the world's surface in alternation side by side and are represented
> by the continents and the ocean floors, respectively. It is therefore very surprising that
> scarcely anyone has tried to explain this law, which has, after all, been well known for some
> time.[42]

To Jordan the hypsographical law was solid evidence of an expanding Earth and
ultimately of a decreasing gravitational constant. As early as 1955, shortly before
the publication of the second edition of *Schwerkraft*, Hans C. Joksch, an astronomer
at the Münster University Observatory, analysed by rigorous statistical means the
Earth's hypsographical curve, pointing out that it consisted of three rather than two
levels. In agreement with "the considerations of P. Jordan and J. E. Fisher"
concerning a varying gravitational constant, he suggested that early expansion
had first disrupted the original sialic crust.[43] Subsequent expansion would then
have disrupted a second layer, which was in turn separated by the third oceanic
layer. Joksch's paper was one of the very few references in the contemporary
geological literature to Jordan's theory—and the only one to Fisher. The Swiss-
Austrian geophysicist Adrian Scheidegger described Joksch's theory and its foundation in Jordan's "highly speculative" suggestion of a decreasing gravitational
constant.[44]

Jordan also referred to other of Fisher's ideas, among them that the orogenic
forces responsible for the formation of mountains were due to the expansion of the
Earth and hence could be understood on the basis of the varying-gravity hypothesis.
Moreover, Fisher had argued that the Earth must cool as a result of its adiabatic

---

[41] Jordan (1962b), p. 287. Sima, an older name derived from silicon and magnesium, denotes the
lower layer of the Earth's crust composed of basaltic rock. It lies below the granitic shell that forms
the foundation of the continental masses and is known as sial (from silicon and aluminum).

[42] Wegener (1966), p. 37, an English translation of the fourth edition of Wegener's classic *Die
Entstehung der Kontinente und Ozeane* first published in 1915.

[43] Joksch (1955).

[44] Scheidegger (1958), p. 9 and p. 157.

expansion, thereby balancing the heat produced by radioactive minerals. Jordan considered it "an attractive thought."[45] Some of Jordan's arguments related to the consequences the $G(t)$ hypothesis would have on the structure and history of the Moon. One of the consequences was a variation in the Moon-Earth distance and another was the cracks or rills on the surface of the Moon mentioned in Jordan's letter to Pauli quoted above. Jordan first mentioned the rills as possible evidence for expansion in 1955 and later in greater detail, concluding that "there is nothing in our observations of the lunar rills to contradict our interpretation of them as consequences of Dirac's hypothesis."[46] From the size and number of rills he estimated an average increase in the Moon's radius of about 0.005 mm per year.[47]

The implications for the history of the Moon of a decreasing gravitational constant was studied by Horst Gerstenkorn, a high-school teacher and accomplished amateur astronomer in Hannover who made important contributions to lunar theory.[48] The Dirac–Jordan variation of gravity would have caused tidal friction to be stronger in the past. Gerstenkorn showed that, as a consequence, the Moon must originally have been a dwarf planet that was captured by the Earth's gravitational field only 600–800 million years ago. The age of our satellite would thus be much smaller than what was ordinarily believed (namely, nearly the same as the age of the Earth or about 4.5 billion years). The German amateur astronomer found the scenario to be doubtful and the $G \sim 1/t$ assumption to be "very hypothetical." Jordan noted the problem, but apparently without considering it a serious one.[49]

Volcanism was yet another phenomenon that Jordan thought was the result, at least in part, of tensions in the crust and ultimately explainable by a diminishing gravitation. He tended to believe that volcanoes were a special feature of the Earth, not to be found in other planets with the possible exception of Venus. This idea he credited to Hans-Jost Binge, a young student of his who had written a manuscript on volcanoes and intrusions that Jordan found to be very convincing.[50] Binge had suffered since his birth from cerebral palsy and was unable to communicate normally by either speech or writing. His manuscripts were all written by his mother, who alone was able to understand him. Binge died at the age of forty, shortly after having graduated with a dissertation on the geophysical consequences of the $G(t)$ hypothesis.[51] Jordan wrote movingly of Binge as "a great talent [who] was subjected to a tragic fate."[52]

---

[45] Jordan (1955), p. 228.

[46] Jordan (1971, p. 47, 1955, p. 237).

[47] Jordan (1971), p. 95. Jordan also believed that the (in)famous "canals" on Mars were real and that they were due to expansion of the planet. Jordan (1961b), p. 20.

[48] Gerstenkorn (1957). See also Brush (1996c), pp. 200–202.

[49] Jordan et al. (1964), p. 513, Jordan (1971), p. 106.

[50] Binge (1955), Jordan (1955, p. 233, 1959b, p. 783, 1973, pp. 68–69).

[51] Binge (1962), Jordan (1971), pp. 120–121.

[52] See Jordan (1966), p. 100 and also Jordan (1971), pp. 118.

Despite his severe handicap, Binge was able to write a couple of papers on astrophysical subjects and also a note on his theory of volcanism.[53] According to the Binge–Jordan view, part of the material of the Earth's crust was under a pressure lower than that at the time of its formation and would therefore be in an unstable state. Volcanic activity consisted of explosive eruptions that originated from phase transitions in the material and were connected with folding processes. The only way to account for the diminishing pressure causing the explosions was to assume a systematically weakening gravitational constant. Indeed, Jordan argued that volcanism was "a testimony to the Dirac hypothesis of a decrease in the force of gravity."[54]

As mentioned in Sect. 2.3, Jordan did not accept Teller's paleoclimatological argument against Dirac's hypothesis. Still in 1952, in the first edition of *Schwerkraft*, he seems to have been unaware of ter Haar's cloud hypothesis which came to work as Jordan's preferred protection against Teller's conclusion. By the mid-1950s Jordan considered a warmer Earth in the past to be in agreement with Dirac's hypothesis if only supplemented with the idea of a Venus-like Earth in the Carboniferous period. This idea he considered to be "confirmed in an extremely convincing manner" by the type of vegetation in the Carboniferous which strongly indicated a total coverage by clouds.[55]

Jordan also considered the apparent conflict with older ice ages predicted by the astronomical theory proposed by the Serbian mathematician and astronomer Milutin Milanković, but only to conclude that there was no real conflict. On the contrary, Jordan suggested that the $G(t)$ hypothesis was "able to answer the question, why the inevitable climatic changes controlled astronomically could lead to glaciation during the last 10^6 years, but not in former times."[56] He referred to the increased solar constant in the past, which he estimated to be just enough to prevent the icing that would otherwise follow the Milanković cycles.[57]

In short, Jordan was convinced that he had vindicated Dirac's hypothesis empirically. Expansion of the Earth was real and it was mainly due to a gravitational constant decreasing inversely with the age of the universe. Jordan spread the gospel of the gravity-induced expanding Earth at every possible occasion, at conferences, in public lectures, and in a variety of scientific and popular publications. Thus, in a biography of Einstein published in 1969, he included a lengthy section on these non-Einsteinian topics.[58] By the early 1960s Jordan seems to have believed that his and others' ideas of an expanding Earth was about to win general recognition among geophysicists. In his unpublished ARL report of 1961 he wrote:

---

[53] For Binge's contributions to astrophysics, see *Zeitschrift für Naturforschung* A **6** (1951): 49–53, **7** (1952): 440–444, and **11** (1956): 874.

[54] Jordan (1971, p. 123, 1962b, p. 286).

[55] Jordan (1962b, p. 285, 1967).

[56] Jordan (1964), p. 115.

[57] Jordan et al. (1964), pp. 516–518.

[58] Jordan (1969b).

"The opinions of leading specialists in this field are converging now to acknowledge expansion of the earth as an empirically stated fact; differences of opinion remain only concerning the amount of this expansion."[59] But Jordan's optimism was unfounded.

In the mature version of his theory Jordan presented the separation of South America from Africa as purely a result of global expansion at a rate about 5 mm per year.[60] However, according to Jordan, the expansion of the Earth had taken place in two stages, for up to the end of the Paleozoic the rate had only been 0.5 mm per year. He argued that the faster expansion rate of

$$\frac{1}{R}\frac{dR}{dt} \cong 10^{-9} \text{year}^{-1}, \quad \frac{dR}{dt} \cong 5 \text{ mm year}^{-1}$$

was characteristic for the later eras. Contrary to some other expansionists (including Egyed and Dicke) Jordan was convinced that expansion made continental drift in the sense of Wegener, superfluous. He estimated that about 90 % of the expansion was due to decreasing gravity, stating that "Dirac's hypothesis allows us to understand earth expansion as a *sufficient* cause to draw the different parts of the continental-block-layer from each other."[61]

The combination of geophysical, astronomical and astrophysical evidence showed to Jordan's satisfaction that "Dirac's hypothesis should not really be considered as a hypothesis, but rather as empirically proven knowledge."[62] A few years later he repeated that the $G(t)$ hypothesis provided "an empirical justification for the Earth's expansion which is independent of any theoretical speculation."[63] As the hypothesis of a decreasing G justified the expanding Earth, so did "our present knowledge of the earth ... make[s] the correctness of Dirac's hypothesis an established fact."[64] Although Jordan realized that the evidence was indirect and in need of further support from astronomical measurements, such as radar determinations of the distance to the Moon and the planets, he nonetheless accepted it as proof. The German physicist Wolfgang Kundt recalled about his former professor that his confidence in the $G(t)$ hypothesis was stronger than in the reality of conflicting evidence.[65] On the other hand, Jordan was far more open to conflicting evidence than Dirac, who simply tended to disregard them.

---

[59] Jordan (1961b), p. 16.

[60] Jordan (1971), p. 92.

[61] Jordan et al. (1964), p. 508, Jordan (1961b). Emphasis added.

[62] Jordan (1959b), p. 795.

[63] Jordan (1963), p. 282.

[64] Jordan (1962a), p. 600.

[65] Kundt (2007).

Jordan was a professed positivist, anti-materialist and empiricist.[66] In the late 1930s he argued that Dirac's theory based on the LNH was in reality an "empirical cosmology" solidly based on and justified by hypothesis-free observational facts.[67] Twenty years later he adopted the very same methodology, that is, to infer the truth of a theory by the facts it can explain. In his writings from the period he referred several times to Wegener's theory of drifting continents. Although he admired Wegener, he argued that his theory was defective and inferior to his own view of an expanding Earth. Yet, methodologically his approach was strikingly similar to Wegener's, namely, to consider a large and diverse group of unexplained or poorly understood geological phenomena and then argue that they could best be explained by a general hypothesis, in his case a diminishing gravitational constant. Jordan expressed his method and conclusion as follows:

> Many facts, which up till now could only be interpreted by ad hoc hypothesis, can be understood if one assumes Dirac's hypothesis—the one hypothesis replaces numerous others. This is true to such an extent that it is almost suspicious on the face of it; almost every problem which, so far, had received no satisfactory solution could form a basis for Dirac's hypothesis. On reflection we realise that this is the overall result to be expected if the hypothesis is valid.[68]

Referring to his book *The Expanding Earth*, in more general terms he spelled out his methodological strategy in the following way[69]:

> It is the explicit purpose of the book to examine whether Dirac's hypothesis can be (a) disproved, (b) proved correct, or (c) left as an open question, from the empirical data available. It would be disproved if we were to draw from this hypothesis a conclusion which is contradicted empirically. We could obtain proof, if certain facts were to be found which are only explicable by the hypothesis, and if a complete survey would show that no contradiction exists. For each range of phenomena which can be tested in this way we must first of all attempt to state all the observed facts freed from any hypothetical interpretation.

The belief that pure facts of nature can be stated unambiguously without involving any theoretical assumptions is not shared by philosophers, but it was an integral part of Jordan's own philosophy of science. In his very first paper dealing with Dirac's hypothesis he stressed that it was possible "to distinguish quite clearly between what are *observational facts*—and as such *independent of any theory*—and what are the results of theoretical considerations."[70] The explanations Jordan offered were qualitative and not very detailed, and they all derived from the $G(t)$ hypothesis rather than specifically from his extended theory of general relativity.

---

[66] See Jordan (1934), an essay on the positivist concept of reality in which he concluded that "the method of positivism is nothing but the scientific method in its purest form."

[67] Jordan (1938).

[68] Jordan (1971), p. 15.

[69] Jordan (1971), p. 19.

[70] Jordan (1937), p. 515.

In other words, one could adopt the $G(t)$ hypothesis without accepting his scalar–tensor theory, which is what most scientists in favour of the hypothesis did.

Another feature in much of Jordan's work in the decade from 1955 to 1964 deserves mention, namely, his use of sources and lack of specific references to the scientific literature. For one thing, he often referred to authors whose work could not be looked up because they had published nothing. In other cases he referred to correspondence or unpublished lectures and manuscripts. Moreover, he did nothing to distinguish between orthodox and unorthodox ideas, or between ideas suggested by amateurs and by reputed earth scientists. Jordan typically mentioned numerous names in his articles and books, but only in very few cases did he connect the names with references to the literature; and when he did, his selection was quite arbitrary. His 1962 article in *Reviews of Modern Physics* contained about forty names and not a single reference to works by the mentioned authors.[71]

One may reasonably assume that Jordan's way of writing diminished his reputation as a serious scientist among many geologists and geophysicists in the British-American tradition. Only in 1966, in a major monograph translated into English five years later as *The Expanding Earth*, did he present his theory in full and supplied it with a comprehensive list of references. By that time the pendulum had swung towards global plate tectonics and the expanding Earth become a minority view. Although the book was often referred to, it made very little impact on the development of geophysics.

In evaluating Jordan's work in the earth sciences it should be kept in mind that he was busy with many other issues in the post-World War II period. Thus, from 1957 to 1961 he was a member of the *Bundestag*, the West-German federal parliament, for the Christian-Democratic party. In addition to his political work, Jordan was also much occupied with public lectures, popularizations of science and generally with acting as what in German is known as a *Kulturträger*, an emissary of culture.[72] On the scientific side, he and his collaborators in Hamburg did much work in general relativity and also in pure mathematics, including the theory of algebra and groups.

Jordan's publications on varying gravitation and the expanding Earth were known by physicists and geophysicists but without attracting a great deal of attention. Paul Wesson found Jordan's theory of the Earth to be interesting but also "dubious from the geophysical aspect."[73] To a large extent Jordan's theories were overshadowed by those of Robert Dicke, who published in English and in the form of papers in widely circulated journals. Jordan, on the other hand, mostly wrote in German and summarized his work in monographs of which only one

---

[71] Jordan (1962a). On the other hand, in a German review article from the same time he did add a list of references. See Jordan (1961a).

[72] Beyler (1994), pp. 485–495.

[73] Wesson (1973), p. 25.

appeared in an English translation.[74] Several of his publications on the extended
relativity theory, the expanding Earth and the $G(t)$ hypothesis appeared in the not
widely read proceedings of the Academy of Sciences and Literature in Mainz.
Jordan was a co-founder of the Academy in 1949 and from 1963 to 1967 he served
as its president. Among the few German geologists with whom Jordan interacted
was Harm Glashoff at Mainz University "who has given me much friendly advice."
In his monograph on the expanding Earth Jordan emphasised that "a careful study
of his [Glashoff's] work seems essential for any serious specialist's discussion of
the geological consequences of Dirac's hypothesis."[75]

When *Die Expansion der Erde* (The Expansion of the Earth) appeared in English
in 1971, continental drift in the form of plate tectonics had become the preferred
theory of the development of the Earth's surface. Another book published the same
year announced the death of the expanding Earth theory as follows: "The idea of an
expanding Earth has a long and interesting history, but recent observations have
shown that any expansion is unlikely or so extremely small that it is not significant
during at least the last 1000 million years."[76] According to the authors, two geophysicists at the University of Newcastle, non-expansion was proved primarily by
magnetic measurements of the Earth's paleoradius and determinations of the
number of days in the year by means of the coral growth rings method. We shall
return to the two methods in Sect. 4.3.

What matters is that by the early 1970s the expanding-Earth theory was no
longer considered a serious alternative to plate tectonics by the majority of earth
scientists. As noted by Homer Le Grand, "When in 1971 his monograph on earth
expansion was translated into English, it was not only out-of-date in terms of its
geology but it had been overtaken by the emergence of newer versions of Drift
including plate tectonics."[77] Whether for this reason or not, *The Expanding Earth*
received no review in leading and widely circulated science journals such as *Nature*
or *Science*.[78] Jordan's mostly cited paper received 51 citations until 1980, whereas
the one in the leading physics journal *Reviews of Modern Physics* was only cited

---

[74] Menard (1986), p. 144 and p. 316 erroneously states that Jordan's book on the expanding Earth
dates from 1952 and was translated into English in 1966, which he considers an indication of "the
intensity of interest in the expansion hypothesis during the plate tectonic revolution." He evidently
mixed up the first edition of *Schwerkraft* and Jordan's later *Die Expansion der Erde*. In 1966 there
was indeed a great deal of interest in the expanding Earth hypothesis, but Jordan's book attracted
only very limited attention.

[75] Jordan (1971), p. 145. Jordan referred to a work on the dynamics of the Earth from the viewpoint
of the $G(t)$ hypothesis that Glashoff had published in the proceedings of the Mainz Academy. See
Glashoff (1966).

[76] Tarling and Tarling (1971), p. 84.

[77] Le Grand (1988, p. 227). For more on the late phase of Earth expansionism, see Sect. 4.4.

[78] The German 1966 edition was reviewed in *Tectonophysics* **4** (1967): 117–120 by Hans Georg
Wunderlich, a professor of geology at Göttingen University. Although *Tectonophysics* was an
English-language journal, the review was in German. The English translation of Jordan's book was
reviewed by the Manchester astronomer Michael Moutsoulas in *Geoexploration* **11** (1973):
197–198.

twelve times in the same period.[79] The numbers of citations were much smaller than for Dicke and his collaborators. As Jordan's extensive work on the expanding Earth received but limited attention in the 1960s, so it is considered of marginal interest (if any at all) to modern historians dealing with the recent revolution in the earth sciences. It may be an exaggeration to say that Jordan has been written out of the history of the earth sciences, but if so it is a slight exaggeration only.[80]

3.3   Dicke and the Earth Sciences

As is evident from his Chapel Hill address and the two subsequent papers in *Reviews of Modern Physics*, by 1957 Dicke had obtained solid knowledge of a series of geophysical subjects, which he discussed expertly (Fig. 3.4). Some of the knowledge he got from discussions with his Princeton colleague Harry Hammond Hess, who had joined the faculty in 1934 and in 1950 was made head of Princeton University's Department of Geology. In papers from the early 1960s Dicke acknowledged "the many suggestions and ideas I have derived from conversations with Professor H. Hess of the Princeton geology department."[81] In 1960 Hess formulated the crucial idea of "sea floor spreading" and thereby made a seminal contribution to what would soon become known as plate tectonics. According to Hess' hypothesis the sea floor was generated at mid-oceanic ridges by the convection of the Earth's mantle, and from there it spread. Hess originally circulated the sea floor spreading hypothesis by means of a preprint only, whereas the first publication on the subject was due to the American marine geologist Robert

Fig. 3.4 Robert Henry Dicke. Reproduced with the permission of the American Institute of Physics

---

[79] Web of Science data. The two papers are Jordan (1959a, 1962a).

[80] Frankel (2012b), pp. 278–354 deals extensively and scholarly with Earth expansionism without mentioning Jordan or his books. Jordan is also absent from the review of expanding Earth theories in Nunan (1998). Jordan's book of 1966 appears in the bibliography of Sudiro (2014), but there is no mention of Jordan in the review article itself. The same is the case with Nunan (1988).

[81] Dicke (1962a), p. 664 and Dicke (1961b), p. 106.

Dietz, at the U.S. Coast and Geodetic Survey, who independently had arrived at the same idea. It was also Dietz who coined the name. Nonetheless, Hess is generally credited with first proposing sea floor spreading.[82]

In an interview of 1975, Dicke said:

> Long before the average geologist in the country took this continental drift, and plate tectonics, to mean anything at all—Harry Hess, over in our geology department, had a clear picture of what was going on. Anyway, it's obvious that as gravitation is getting weaker, the earth should expand slightly, and I noticed in my readings at that time that there were cracks in the mid-Atlantic ridge in the ocean, oceanic cracks. So this suggested that these cracks might be the result of tension, due to the earth expanding . . . . I went over and talked to Hess about this. We laid out a beautiful picture of the Atlantic Ocean crust moving, and trenches, and island arcs, and all the—the whole plate tectonic game was laid out for me, and this was the late fifties.[83]

On the suggestion of Hess, Dicke visited Yale University in 1959 to discuss with Warren Carey the evidence for an expanding Earth and its relation to decreasing gravity. Carey, who stayed as visiting professor at Yale, had at the time become convinced that the present surface of the Earth was the result of a long phase of expansion. In late 1959 and early 1960 he gave several lectures in Princeton.[84]

Dicke also collaborated with several physics colleagues and graduate students on geo- and astrophysical problems related to the $G(t)$ hypothesis. One of them was William Jason Morgan, who in 1964 wrote his physics Ph.D. thesis under Dicke and with Hess on his committee. The title of the thesis was "An Astronomical and Geophysical Search for Scalar Gravitational waves." Such "φ-waves," as they were also called, were expected from the Brans–Dicke scalar–tensor theory and Dicke wanted to know if they actually existed and what observable effects they might have.[85] One possibility was that the φ-waves, which were thought to be caused by a variation of G in space and time, triggered earthquakes. This was the subject that 26-year-old Morgan investigated in a paper co-authored by J. O. Stoner and Dicke.[86] Analysing nearly 2000 earthquakes in the period from 1904 to 1952 with the purpose of finding periodicities, the three authors concluded that there existed a statistically significant annual periodicity possibly due to a corresponding variation in the gravitational constant. The mechanism might be that the change in G led to periodic stress in the Earth and that the stress triggered earthquakes. However, Morgan and his two co-authors cautiously pointed out that the observed annual period could not be interpreted as strong evidence for the hypothesis of a varying G.

---

[82] For details, see Frankel (2012c), pp. 280–319.

[83] Interview by Spencer Weart of 18 November 1975, the Niels Bohr Library and Archives, the American Institute of Physics. See http://www.aip.org/history/ohilist/31508.html

[84] Carey (1988), p. 119 and p. 141.

[85] Dicke (1964c).

[86] Morgan et al. (1961). Morgan and Stoner were at the time National Science Foundation pre-doctoral fellows working with Dicke's gravity group in Princeton. For a later attempt to link the periodicity of earthquakes to the variation of G, see De Sabatta and Rizzati (1977).

At the time Morgan had taken no course in either geology or geophysics, but his work with the thesis turned him toward research in geophysics. At the end of 1964 he was hired by the German-American geophysicist Walter Elsasser, who had joined Hess in Princeton in 1963.[87] Four years later, in what was only his fifth paper, Morgan presented a cornerstone of the new plate tectonics in the form of a quantitative, mathematically formulated theory of trenches and faults. When Morgan received the prestigious National Medal of Science in 2003 he noted that Dicke had received the same award in 1970. He paid the following tribute to his former thesis adviser: "My apprenticeship with him more than 40 years ago was where I learned what science is—how to formulate and attack a scientific problem. His approach and attitude toward science remain with me today."[88]

Dicke was not originally interested in either geology or geophysics. But, as an undergraduate at Princeton, he wanted to take at least one course in each of the major sciences. From this honourable intention, he made one exception and that was geology.[89] By 1957 he had become seriously interested in the subject, as indicated by his membership in the American Geophysical Union. Generally Dicke had easy access to several of the American geophysicists and oceanographers who were instrumental in the plate tectonic revolution. Jordan in Hamburg did not have the same advantage and neither, perhaps, the same interest.

In two broad-ranging papers of 1957 and 1962 Dicke surveyed how a varying gravitational constant as given by either Dirac's theory or the slower version of the scalar–tensor theory would affect the Earth and the Moon. His aim was not primarily to contribute to the geological sciences, but rather to use geological and astronomical data as tests for the $G(t)$ hypothesis and more specifically for the Brans–Dicke theory of gravitation. He hoped in this way to throw light on how "physics, astrophysics, and geophysics may become completely enmeshed in a gravitational study."[90] This synoptic view was characteristic for much of Dicke's research in the period.

The phenomena that Dicke dealt with were largely the same as those Jordan had covered. Moreover, the two physicists shared the same "Wegenerian" method, namely, to investigate within a synthetic perspective a wide range of geophysical problems that were not yet well understood. If they could be better explained on the assumption of varying gravitation the hypothesis would gain in credibility, perhaps even come out as empirically justified. Wegener's arguments for continental drift did not refer only to geology, geophysics and geodesy, but also to astronomy,

---

[87] Frankel (2012d), pp. 474–475. Elsasser, who originally worked in atomic and quantum theory, had known Dirac since the late 1920s. In an appendix to a paper of 1971 Elsasser dealt with the geological effects of Dirac's $G(t)$ hypothesis which "appeared a few decades after [before?] the confidence of earth scientists in the ancient contraction ('shrinking apple') model of mountain building had begun to be shaken." Although Elsasser found the hypothesis of gravity-driven Earth expansion to be interesting, he did not support it. See Elsasser (1971).

[88] Schultz (2003).

[89] AIP interview of 18 June 1985. http://www.aip.org/history/ohilist/4572.html

[90] Dicke (1961c), p. 797.

paleoclimatology and zoology. They appealed to the coherence of the broader, global picture but invited criticism from the point of view of the special sciences.[91] The same was the case with Jordan's arguments, and to a lesser extent with Dicke's, for the expanding Earth.

Although the results obtained by Dicke did not differ substantially from Jordan's, his methods and conclusions did. While Jordan was convinced that evidence based on the study of the Earth amounted to proof of the $G(t)$ hypothesis, Dicke's conclusion was more reserved. None of the evidence "can be used to give strong support to Dirac's hypothesis," he cautiously stated in his 1957 paper, adding that a variation of the gravitational constant could not be excluded. Five years later, he concluded that his analysis of examples from geophysics "cannot be marshaled as evidence for a gradual decrease in the gravitational constant."[92] To him, the case for $G(t)$ was neither proven nor disproven. The two physicists also had different attitudes to the expanding Earth, which Jordan was strongly committed to. Dicke, on the other hand, only supported the hypothesis half-heartedly and for a rather brief period of time (see below).

In one of his 1957 papers in *Reviews of Modern Physics*, Dicke included a discussion of the consequences that the $G(t)$ hypothesis would have for the formation and surface of the Moon. If the Moon had originally been formed in a molten state and subsequently solidified—Dicke assumed it to have occurred 3.25 billion years ago—its surface area would since then have increased slightly because of the expansion caused by decreasing gravity. Like Jordan, he pointed to the rills of the Moon as possible evidence and, also like Jordan, he emphasized that the expansion effect would have been much smaller than for the Earth. The problem of the formation of the Moon was at the time unresolved, with a capture hypothesis and a rival fission hypothesis being the most popular candidates.[93] Dicke argued that the hotter Sun in the past, as predicted by the $G(t)$ hypothesis, supported the view that the Earth and the Moon were formed as molten bodies. Tidal waves in a liquid Earth might have produced the mechanism for fission in accordance with the idea originally suggested in the late nineteenth century by the British mathematician and astronomer George Howard Darwin, a son of the famous naturalist Charles Darwin.

The increasing interest in Brans–Dicke cosmology led G. Shahiv and John Bahcall to investigate the effect of $G(t)$ on evolution of the Sun, thus extending the line of research started by Pochoda and Schwarzschild a few years earlier.[94] While in the classical model (G = constant; $\omega \to \infty$) the solar luminosity increases mildly with time, Shahiv and Bahcall found that for smaller values of ω it changes to a decrease. The more rapid depletion of hydrogen caused by a higher luminosity in the past would influence the neutrino flux, and Shahiv and Bahcall concluded that

---

[91] Le Grand (1988), pp. 80–96, Frankel (1976).

[92] Dicke (1962a), p. 664. See also Dicke (1964b), p. 173.

[93] On theories of the formation of the Moon, see the careful account in Brush (1996c).

[94] Shahiv and Bahcall (1969).

in the Brans–Dicke model with $\omega = 5$ it would be about twice as great as the one calculated from the then standard theory based on the assumption of constant G. Since the solar neutrino flux had recently been measured by Raymond Davis in rough agreement with the latter assumption, Shahiv and Bahcall concluded that their result constituted a problem for the Brans–Dicke theory.

"It is remarkable that the rate of heat flow from the earth should present a crucial test for a physical hypothesis," Dicke stated, referring to the $G(t)$ hypothesis.[95] Having discussed the problem of heat flow from the interior of the Earth in a preliminary way, he handed it over to one of his students, Charles T. Murphy, who wrote an undergraduate thesis on it. In an interview of 1975 Dicke recalled: "I had one student, for example, look at the heat flow problem from this point of view [decreasing gravity], because . . . if you have the interior hot, as we understand it is, and if the temperature curve for the mantle of the earth is near the melting point, if you lower the pressure inside, . . . you can calculate the way in which heat flows out this way. And it agrees rather well with what is observed, to a factor of two."[96]

A few years later an extended version of Murphy's thesis was transformed into a joint paper with Dicke published by the American Philosophical Society.[97] Based upon considerations of the energy sources of the Earth, Dicke and his co-author inferred that radioactivity and other known processes were not enough to account for the observed heat production of about 50 erg s^{-1} cm^{-2} reaching the Earth's surface area. The additional energy source, they suggested, might have its roots in gravity weakening in time. The idea was, roughly, that as G decreases, the pressure and melting point of the mantle will decrease and the interior of the Earth cool. Heat will flow out of the core and mantle at a rate greater than if G were constant.[98] Moreover, the induced heat flow added to the heat generated by radioactivity makes it possible for convection currents to occur in the mantle.

The net result of a series of complex mechanisms was this: "The decreasing gravitational constant makes available the internal heat of the earth as a steady-state convective system in the mantle and makes convection a more likely possibility."[99] Calculating the heat release brought about by this effect, Dicke and Murphy found the value 2.5×10^{12} cal s^{-1} for the entire Earth surface, or about 20 erg s^{-1} cm^{-2}. "Thus a large portion of the observed heat flow might originate deep in the earth," they concluded. The proposed mechanism might also explain the "mystery" of the rate of heat flow from the ocean floors, namely, that it was nearly the same as that from the continents despite the much higher content of radioactivity in the latter.

---

[95] Dicke (1957a), p. 361.

[96] AIP interview, http://www.aip.org/history/ohilist/31508.html

[97] Dicke (1962a), p. 660, Murphy and Dicke (1964).

[98] Dicke seems at the time to have been undecided with regard to the physical state of the inner core. In Dicke (1962a, 1964b) he stated that the inner core was solid, but in other of his publications from the same time he expressed doubts about Bullen's evidence for a solid inner core, which he judged to be marginal. The turning point in the acceptance of the solid inner core only occurred in the late 1960s. See Brush (1996a), p. 201.

[99] Murphy and Dicke (1964), p. 243.

Dicke took the mean heat flow from the ocean floor to be 35 erg s^{-1} cm^{-2} and found the effect of a decreasing G to represent almost half this figure.[100]

Richard von Herzen, a geophysicist at the Woods Hole Oceanographic Institution in Massachusetts, was not ready to accept Dicke's "exotic" explanation of heat escaping from the Earth's interior. He suggested that the theory and its basis in the $G(t)$ hypothesis was a "speculative mechanism [that] should be accepted with a small grain of salt ... until a need to accept its implications increases."[101] Other geophysicists and geologists needed more than a small grain of salt to accept the $G(t)$ mechanism.

Murphy and Dicke were well aware that convection in the mantle was a controversial hypothesis and had remained so since the days of Wegener.[102] Indeed, convection currents of radioactive origin were the favoured mechanism for the horizontal movements of the continents and consequently denied by Jeffreys and many other critics of Wegener's theory. Noting the close connection between the convection hypothesis and continental drift, Jordan sharply criticized the hypothesis as contrived and unnecessary. He thought it was inconsistent with the second law of thermodynamics.[103] On the other hand, he praised the work of Murphy and Dicke for offering a new and physically more correct mechanism for mantle convection without claiming that it supported continental drift. According to Jordan, the Murphy–Dicke theory was irrelevant to his own version of a more rapidly expanding Earth. None of the effects of a slowly decreasing G that Murphy and Dicke predicted for the interior of the Earth violated established geophysical facts, but unfortunately the effects were not empirically verified. Consequently, the two authors concluded, "they do not directly verify the [varying-G] hypothesis."

Nor did detailed studies on the rotation of the Earth and the temperature of meteorites succeed in clearly identifying effects of the decreasing gravitational constant. Dicke concluded that the astronomical data allowed a rate of change of G in agreement with the Brans–Dicke theory ($\sim 10^{-11}$ year^{-1}), but he was unable to present his analysis in stronger terms, namely, as positive evidence for the theory. Lack of counterevidence is not evidence.[104] He returned to the issue in an article of 1969, this time with a focus on the geophysical consequences of a changed rotation of the Earth rather than focusing on the varying constant of gravitation.[105]

The situation was no more promising in an investigation Dicke made with his former student James (or Jim) Peebles concerning the amount of argon in

---

[100] Dicke (1962a). For the Earth as a heat engine, see also Scheidegger (1958), pp. 55–57, who quoted the average heat flow to be about 1.2×10^{-6} cal s^{-1} cm^{-2} or 50 erg s^{-1} cm^{-2}.

[101] Von Herzen (1967), p. 213.

[102] On the problem of mantle convection, see Oldroyd (1996), pp. 255–257, and Le Grand (1988), pp. 112–117.

[103] Jordan (1966), pp. 80–83, Jordan et al. (1964), p. 507.

[104] Dicke (1966).

[105] Dicke (1969).

meteorites.[106] Peebles had come to Princeton in 1958 to study particle physics but soon became part of Dicke's gravity group. On the suggestion of Dicke and motivated by his fascination with Mach's principle, Peebles wrote in 1962 his dissertation on varying constants of nature. The title was "Observational Tests and Theoretical Problems Relating to the Conjecture that the Strength of the Electromagnetic Interaction May Be a Variable." In a follow-up paper Peebles discussed the possibility of a varying fine structure constant in relation to the isotropy of space, which was also the subject of a joint paper by Peebles and Dicke.[107] Some of the papers co-authored by Dicke and Peebles in the 1960s relied in part on Peebles' dissertation. About his encounter with Dicke and his group, Peebles recalled:

> I was fascinated by the variety of topics under discussion and intimidated by how much everybody knew. Dicke, in particular, . . . was drawing from a deeper well of understanding of the physics of the real world than anyone else I have encountered. . . . Bob's motivation was his fascination with Mach's principle, which might be read to say that as the universe evolves so do the laws of physics. I was fascinated by all the evidence one could bring to bear, from the laboratory to geology and astronomy. My evident lack of interest in Mach didn't seem to bother Bob.[108]

The idea behind the Peebles–Dicke meteorite study was that a higher temperature in the early history of the solar system would have caused an anomalous loss of argon from meteorites. But the method only made it possible to estimate an upper value for the relative G-dependence on time of 10^{-10} per year. When Dicke and Peebles reviewed the empirical arguments for the Brans–Dicke theory, the words "not compelling" appeared repeatedly.[109] It may have been this state of affairs, namely, the inadequacy of astronomical, geological and geophysical tests to yield unambiguous answers that caused Dicke to withdraw from geophysics. For a few more years he continued doing work in the area, but then abandoned it. His last contribution to geophysics dates from 1969.

According to Dicke, a weakening gravity and an expansion of the Earth might play some role in terrestrial history, but

> . . . it just seemed to me to be so deeply buried in all the other things that it would be hard to separate it out, in an unambiguous way. . . . I decided after a while that it was just too hard to try to get fundamental physics out of the earth.[110]

There were other reasons as well for Dicke's retreat from geophysics, not least that his interest in cosmology was boosted by the discovery in 1965 of cosmic

---

[106] Peebles and Dicke (1962a).

[107] Peebles (1962), Peebles and Dicke (1962c).

[108] Peebles et al. (2009), p. 185.

[109] Dicke and Peebles (1965).

[110] Interview by Spencer Weart of 18 November 1975, the Niels Bohr Library and Archives, the American Institute of Physics. http://www.aip.org/history/ohilist/31508.html. Freeman Dyson echoed Dicke's sentiment. "Until geophysics becomes an exact science," he wrote, it would be impossible to relate the past temperature of the Earth or the heat flow through the Earth's crust in an unambiguous way to a decrease in G. See Dyson (1972), p. 230.

microwave radiation. Apart from engaging fully in the new big-bang cosmological theory, he also turned to the astronomical consequences of the Brans–Dicke theory, in particular the tricky problem of measuring and explaining the assumed oblate shape of the Sun. The precise shape of the Sun affects the value of Mercury's perihelion advance and thus might help to establish existence of the φ field in accordance with the Brans–Dicke theory. However, although the initial measurements of Dicke and his collaborator H. Mark Goldenberg came out in favour of the Brans–Dicke theory with $\omega \cong 6$, they were received with scepticism and not generally accepted.[111] Later measurements from around 1980 strongly supported the Einstein theory of gravitation rather than the rival theory of Brans and Dicke. These measurements could only be explained on the scalar–tensor theory if $\omega \geq 500$, where the Brans–Dicke theory approximates general relativity very closely.

Despite all the annoying uncertainty, according to Dicke there were good reasons to believe that the gravitational constant decreased at a rate of about 3×10^{-11} year^{-1}. If this were the case, the Earth would expand, but only at a modest rate corresponding to an increase in radius R of 0.5 cm per century or 0.05 mm per year.[112] In his paper with Murphy, Dicke concluded that the expansion rate was 0.047 mm per year or "about 110 miles over the last 4 billion years." This was an expansion rate much lower than the one suggested on the basis of empirical evidence by Carey, Egyed and other expansionist geologists. Dicke credited the result of a slow expansion due to decreasing gravity to an unpublished senior thesis written in 1958 by his student G. Hess. Its title was "The Annual Variation of the Length of the Day as Evidence Relating to a Theory of Gravity."[113] The author of the thesis was George B. Hess, a son of the geology professor Harry Hess.

Dicke stated the general relation between the change in G and the resulting change in R as

$$\frac{dR}{R} \cong -0.1 \frac{dG}{G}.$$

Could the system of oceanic ridges and the distributions of the continents be the result of a gravity-induced expansion of the Earth? Not according to Dicke, who emphasized that the required magnitude of the expansion could not be explained in terms of gravity decreasing at a rate of about 10^{-11} per year. The rate would have to be higher by a factor of 100, for which there was no physical justification. Dicke consequently dismissed Carey's suggestion of a strong expansion of this

---

[111] The problem of the Sun's oblateness goes back to 1865, when Simon Newcomb tried to explain the Mercury anomaly that was later solved by Einstein. For an overview of the history, see Rozelot et al. (2010).

[112] Dicke (1957a, 1962a).

[113] Morgan et al. (1961), Dicke (1962a). Copies of Hess' thesis work may no longer be extant. I am grateful to George Hess for having confirmed the authorship of the 1958 thesis (E-mail of 15 January 2015). George Hess took his Ph.D. from Stanford in 1967 on experiments with liquid helium. He subsequently became a professor of physics at the University of Virginia, where he mostly worked in condensed matter physics.

magnitude.[114] In his view, the geological evidence rather pointed towards the continental drift picture with convection in the Earth's mantle as the driving mechanism. As to the effect of a diminishing G, he wrote that "The miniscule effects of a modest expansion would be lost in the magnificent displays produced by convection."[115] This was also his conclusion in a report to the Space Science Board established under the National Academy of Sciences in 1958. According to Dicke's report:

> It is not clear that a general expansion of the Earth is required to cause such a separation [of the Americas and Europe-Africa]. It could also be caused by convection in the mantle. Certainly it must be said that if continental drift has been occurring to the extent indicated by recent paleomagnetic data, the effects of an expansion of radius 47 km per b.y. would be negligible. ... If subcrustal currents are not important, an expansion in radius of only 0.0047 cm per year could produce a medial crack in the Atlantic 2 km wide in only 13 m.y., assuming that half the expansion takes place in the Atlantic.[116]

Dicke obviously was not an expansionist in the sense of Jordan, but neither was he strongly committed to the revived theory of drifting continents. He just considered it the best of the available pictures of the Earth and its history.

In a letter of July 1966 Jordan informed Dicke about his forthcoming book on the expanding Earth and the large value (10^{-9} per year) he found for the rate of decrease of gravity. Jordan planned a visit to the United States and wanted to meet Dicke in Princeton. In his letter of reply Dicke agreed that, "the implications for geophysics and astrophysics of a time rate of change of the gravitational interaction is one of the most fascinating questions that one could consider." But he was sceptical with regard to Jordan's high value for the rate of change: "I am curious to know how you could have a time rate of change of gravitation as great as 10^{-9} per year and am looking forward to reading about it in your book."[117]

3.4   Egyed and the New Expansion Theory

In a review article of 1963, the Hungarian geophysicist László Egyed, director of the Geophysical Institute in Budapest, noted that several authors had independently arrived at the idea of an expanding Earth.[118] He has himself been described as "the penultimate independent discoverer" and the one with whom "the expansion hypothesis entered the normal literature of science."[119] Jordan referred repeatedly to Egyed,

---

[114] Dicke (1964b), p. 162.

[115] Dicke (1962a) and similarly in Murphy and Dicke (1964), p. 226: "If substantial continental drift associated with subcrustal currents should occur, it seems likely that the effects of a small expansion would be masked by the larger effects produced by these currents."

[116] Dicke (1961b), p. 105. The abbreviations "m.y." and "b.y." refer to million years and billion years, respectively.

[117] Jordan to Dicke, 2 July 1966, and Dicke to Jordan, 7 July 1966. Robert H. Dicke papers, box 4, folder 4, Department of Rare Books and Special Collections, Princeton University Library. I am grateful to J. Peebles for providing me with copies of the letters.

[118] Egyed (1963). For a brief biographical account, see Meskó (1971).

[119] Menard (1986), p. 144.

whom he generously called "one of the fathers of the theory of the expansion of the Earth" and "one of the most enthusiastic defenders" of the theory.[120] Egyed was nationally as well as internationally recognized for his work in seismology, geo-magnetism and other branches of geophysics. He was instrumental in the formation of the Association of Hungarian Geophysicists in 1954 and in his later years the communist regime allowed him to attend many international congresses on geo-physics. In 1960 Egyed was elected a corresponding member of the Hungarian Academy of Sciences and ten years later, shortly before his death, he became a full member.

Based on paleogeographical and other evidence Egyed concluded in the mid-1950s that the radius of the Earth had been expanding for the past 500 million years by the slow rate of 0.4–0.6 mm per year.[121] He assumed the rate to have been constant throughout the period. Among the evidence which inspired Egyed to his conclusion were estimates of the change of water-covered continents throughout the history of the Earth. It was generally agreed from geological considerations that the mass of sea water had remained roughly constant. Egyed now reasoned that, on the assumption of a contracting Earth, it follows that the areas covered by sea water would have increased over the last 500 million years; on the other hand, the opposite trend would follow from an expanding Earth. According to Henri and Geneviève Termier at the University of Paris, the area had decreased. Using the data collected by the Termier couple and also by the distinguished Russian geologist Nikolai Strakhov, Egyed interpreted it as "evidence for the hypothesis that the Earth has expanded."

Egyed's argument based on the coverage of the Earth's surface by water relied on the assumption that the hypsometric curve, meaning the proportion of land area at various elevations, had stayed constant during geological time. As pointed out by his critics, there was no good reason to maintain the assumption without which the expansion argument would lose its force. A slight increase in thickness of the continents offered a plausible alternative interpretation of the paleogeographical data. Yet another possibility was that the data might be explained by a combination of known processes such as glaciation, mountain building, erosion and ocean rises.[122]

Egyed's main argument was sharply criticized by Robert Dietz according to whom the Hungarian geophysicist had failed to take into consideration polar and glacial effects. Dietz dismissed the underlying idea of a Precambrian universal sialic crust which fragmented due to expansion. "I cannot take the expanding Earth hypothesis of Egyed ... and others seriously," he bluntly stated.[123] Despite various objections, Egyed kept to the expansion hypothesis. The Termier couple did not intervene in the controversy over the expanding Earth except that they offered some

---

[120] Jordan (1966, 1971, p. 49 and p. 66).

[121] Egyed (1956a, b). See also Frankel (2012b), pp. 279–282.

[122] Scheidegger (1958), p. 11, Armstrong (1969), Hallam (1971).

[123] Dietz (1967), p. 236, who referred to Egyed (1961b). He did not comment on Egyed's argument for the expanding Earth in terms of decreasing gravity.

indirect support by concluding that "global palaeogeography does not display any argument *against* Earth expansion."[124]

To account for the expansion in physical terms Egyed adopted a modified version of a theory of the composition of the inner Earth first proposed by the British theoretical geophysicist William H. Ramsey at Manchester University in the late 1940s. According to Ramsey, the Earth was originally formed as a cold solid body, and not as a molten mass. Rather than explaining the difference between core and mantle in terms of chemical composition, he explained the core by a phase transition of silicate compounds such as olivine into a liquid metallic state due to extremely high pressure.[125] The phase change was facilitated by heat from radioactive decay. Ramsey's high-pressure phase change theory, which contradicted the standard picture of the core as consisting mainly of iron and nickel, was controversial and dismissed by many earth scientists. Yet it survived in various modified forms. Jordan initially thought that Ramsey's explanation of the spheres of discontinuity inside the Earth as due to phase differences was valid, but in 1962 he came to the conclusion that this was not the case. He ascribed his change of mind to discussions with Edward Teller, according to whom Ramsey's hypothesis was physically improbable.[126]

Egyed suggested that the inner core was a remnant of the original solar material out of which the Earth was formed. He believed that the minimum density of the inner core was 17 g cm^{-3}, that the density of the outer core was 9–12 g cm^{-3}, and that the density of the mantle ranged between 3 and 6 g cm^{-3}. Egyed further argued that the matter of the inner core slowly and irreversibly transformed into a stable low-density phase, implying a decrease in the mean density of the Earth. Assuming the mass of the Earth to be constant then led him to "the surprising conclusion that the volume of the Earth is steadily increasing." As a result of the expansion a large amount of tectonic energy would be released. Egyed's theory was ambitious and total in scope[127]:

A new conception of dynamic character is given for the internal constitution of the Earth. ... The expansion of the Earth is able to account for the formation of the crust and oceanic basins, the energies of the tectonic forces and earthquakes, the origin of deep-focus earthquakes, the periodicity of geological phenomena, the continental drift and mountain building, and is supported also by paleogeographical data.

Egyed's claim, or the corresponding claim of Jordan and Binge, that Earth expansion was able to account for the formation of mountains, was not generally accepted even by most fellow expansionists. Crustal shortening seemed necessary for the folding-up of mountains and it was hard to see how crustal shortening could be obtained by expansion.[128]

---

[124] Termier and Termier (1969), p. 101. Emphasis added.

[125] Ramsey (1949).

[126] Jordan (1962b), p. 288. On the fate of Ramsey's theory, see Brush (1996a), pp. 209–213 and Doel (1996), pp. 97–98.

[127] Egyed (1957), p. 106 and p. 101.

[128] Scheidegger (1958), pp. 204–205.

By the late 1950s Egyed thought he had found an alternative or supplementary explanation for the expansion in Dirac's cosmological hypothesis of a decreasing gravitational constant, or what he called the "Dirac–Gilbert equation." On this basis he revised his picture of the structure of the Earth and its evolution over time. Egyed now described the infant Earth as follows:

> The original radius of the Earth was about half the present-day one so that the average density could amount to 35 g cm^{-3} instead of the present 5.5 g cm^{-3}. The surface of the outer core, i.e. the Gutenberg–Wiechert discontinuity, must have lain at a depth of 100 to 200 kilometres below the surface of the Earth, for the critical pressure generating the high-pressure phase of the core was reached at that level already, because of the high value of the gravity acceleration, as postulated by the Dirac–Gilbert equation.[129]

Apart from applying the Dirac–Gilbert $G(t)$ hypothesis to the expanding Earth, Egyed also used it to propose a new hypothesis of the origin of the solar system based on the great angular velocity of the Sun in the past. The planetary hypothesis led to an expression for the orbital radii of the planets that roughly agreed with the Titius–Bode law going back to the late eighteenth century.[130] Originally trained in physics and mathematics, Egyed was apparently aware of Dirac's hypothesis in the late 1930s, but "at that time, I shared the doubts of most physicists concerning this hypothesis."[131]

It was only after Gilbert's argument that the $G(t)$ hypothesis agreed with general relativity that Egyed accepted the hypothesis and made it the basis of his view of an expanding Earth. Egyed and also Holmes somewhat uncritically supported Gilbert in his claim that he had proved Dirac's hypothesis to be a corollary of general relativity. In the opinion of the large majority of specialists in general relativity theory, Gilbert's claim was wrong (see also Sect. 2.5). They maintained that a varying G cannot be reconciled with Einstein's theory of gravitation, which is also the current view. Given that Gilbert's theoretical value for the age of the universe was 4×10^9 years it is remarkable that Carey referred to it as "an impressive success."[132] Egyed was in contact with Jordan, who was happy to have found a brother in arms. The noted Hungarian geophysicist, Jordan wrote, "now prefers to believe that Dirac's hypothesis is correct and gives the theoretical explanation of this expansion, which Egyed believes to be an empirically proven fact."[133]

---

[129] Egyed (1960a), p. 253. On Gilbert's claim as argued in Gilbert (1956), see also Sect. 2.3.

[130] Egyed (1960c). According to the Titius–Bode law, often referred to as just Bode's law, the radii of the planets follow a simple relation given by the number of the planet as counted from the Sun. Today the "law" is generally considered a rule or mathematical coincidence with no theoretical foundation. The name relates to two German astronomers, Johann Daniel Titius (1764) and Johann Elert Bode (1772).

[131] Letter to Arthur Holmes of 31 July 1959, quoted in Frankel (2012c), p. 285, where Holmes' favourable evaluation of the expanding Earth and the $G(t)$ hypothesis is documented. See also Holmes (1965), pp. 983–987.

[132] Carey (1976), p. 451. Theoretical physicists seem to have ignored Gilbert's claim. Among the few who responded to it was Wesson (1973), who found parts of Gilbert's reasoning to be "obscure" and even "somewhat perverse."

[133] Jordan (1962b), p. 287.

Table 3.1 Egyed's expanding Earth

T (10^6 years)	Geological epoch	G (10^{-11} N m^2 kg^{-2})	g (m s^{-2})	R (10^4 m)	ΔR (10^4 m)
2000	Orosirian	8.00	16.6	537	100
1000	Tonian	7.34	12.7	587	50
500	Furongian	7.00	11.2	612	25
250	Early Triassic	6.83	10.4	625	12
0	Present	6.67	9.8	637	0

The table is adapted from Stewart (1970), p. 414

In a paper of 1958 Egyed and his Hungarian collaborator Lajos Stegena derived from the Dirac–Gilbert hypothesis and the assumption of the Earth's core mentioned above that the annual increase in the radius of the Earth was at least 0.3 mm, hence "almost identical with the lower limit of radius increase as determined from observations."[134] Two years later, in a popular review of geophysics, Egyed repeated his empirical arguments in favour of the expanding Earth, adding that the theory provided a simpler and more unified picture of the Earth's surface than other theories. He suggested that without Gilbert's alleged proof there would have been no reason to take Dirac's hypothesis seriously[135]:

> On the basis of rather complicated and unclear considerations of a mostly philosophical nature, in 1939 [*sic*] Dirac concluded that the gravitational attraction had continually decreased during the lifetime of the Earth, that is, the gravitational coefficient becomes smaller. Physicists used to exact scientific reasoning received the result with considerable distrust, and that despite that it came from a physicist as famous as Dirac.

At the time Egyed was aware of Carey's independent work on the expansion of the Earth, of which Arthur Holmes had informed him. He also knew about Jordan's work, referring in 1957 for the first time to *Schwerkraft und Weltall*, but without adopting the $G(t)$ mechanism, which he only did the following year. By the early 1960s he called attention to Dicke's somewhat similar theory. As regards the climatological effects of the $G(t)$ hypothesis Egyed pointed out that the larger solar constant in the past would not lead to a correspondingly larger total incident energy from the Sun. The heating effect would to some extent be compensated by the smaller surface area of the ancient Earth. Nonetheless, Egyed found it plausible that the paleoclimate was warmer than at present and also that it was more uniform and moist.[136]

Table 3.1 gives the values for G, the Earth's surface gravity g and its ancient radius R at various times T before the present. The values are based on Egyed's slow expansion rate $dR/dt = 0.5$ mm per year and Dirac's $dG/Gdt = 10^{-10}$ per year; they assume a constant mass of the Earth.

Contrary to some other expansionists, Egyed did not consider the expansion hypothesis a proper alternative to the idea of drifting continents. On the contrary,

---

[134] Egyed and Stegena (1958).

[135] Egyed (1965), p. 100. Translation of Hungarian paper originally published in 1960.

[136] Egyed (1961a).

in his view it provided a partial explanation of drift or spread that avoided the controversial notion of continents moving with respect to the mantle: "In case the Earth is expanding, continental drift is nothing more than the formation of new ocean basins along the gaping rifts which come to exist between continents."[137] At the 1967 Newcastle conference Egyed once again defended the slow expansion hypothesis, now arguing for an average rate of 0.65 ± 0.15 mm per year.[138] He considered Dirac's $G(t)$ to be part of the explanation, but not the only explanation of why the Earth expands. In addition to $G(t)$ he thought that natural radioactivity and phase changes at high pressure played a role. According to Egyed, without these mechanisms expansion would be unable to explain continental disruption.

The dissociation of continental *drift* in the sense of Wegener and continental *spreading* can be found in several advocates of the expanding Earth. The topology of the Earth, wrote Rhodes Fairbridge, "is not the product of continents floating apart on a globe of fixed radius, but the growth of a new oceanic crust, the continents remaining more or less in their same attitudes vis-à-vis each other, but merely further apart or somewhat rotated."[139] Fairbridge suggested that there had been an early and rapid Mesozoic expansion some 200 million years ago. The expansion still continued today, but at a slower rate comparable to the one argued by Egyed. While favouring the expansion of the Earth, contrary to Egyed and Jordan he did not support the $G(t)$ hypothesis, which he thought was inconsistent with evidence from paleoclimatology.[140]

Many of the earth scientists in favour of the expanding Earth simply ignored the $G(t)$ explanation. Among those who considered it, some ruled it out as unnecessary and extravagant while others dismissed it for empirical reasons, such as Fairbridge did. Elena Alexandrovna Lubimova, a geophysicist at the Academy of Sciences in Moscow, argued that a modest expansion was a natural consequence of the thermal evolution of the Earth. "There is no necessity to involve some speculative theory, connected with the variation of universal constant," she wrote.[141] According to the Russian geophysicist the expansion rate was originally, shortly after the formation of the Earth, 7×10^{-3} cm year^{-1} and had at present decreased to 3×10^{-3} cm year^{-1}. During the first billion years the radius of the Earth had increased by only 50–100 km. The view of a decreasing rather than an increasing slow expansion was unusual and contrary to, for example, the ideas of Carey. Yet Lubimova's reasoning and result received support from another expansionist, the Italian geophysicist Giorgio Ranalli according to whom there was "ample evidence

---

[137] Egyed (1960b).

[138] Egyed (1969a).

[139] Fairbridge (1966), p. 143.

[140] Fairbridge (1964). On Fairbridge's view and its basis in the climate of the ancient Earth, see also Sect. 3.6.

[141] Lubimova (1967), p. 310. Scheidegger (1958), pp. 154–155, admitted the possibility that thermal processes might cause a slight expansion of the Earth but not that such processes were realistic. The question of a thermal expansion of the Earth was also examined by Paul Reitan, a Norwegian geologist, who concluded that temperature changes could have caused an increase in radius of at most 5 km during the last billion years. See Reitan (1960).

that the earth has been subject to expansion."[142] Ranalli briefly considered Dirac's cosmological $G(t)$ hypothesis, which he found to be suggestive but not directly testable.

A tireless advocate of his gravity-based theory of the expanding Earth, Egyed wrote and lectured on it at numerous occasions, not only in Europe but also in Japan and the United States. For example, on 6 February 1961 he presented a paper to the New York Academy of Sciences in which he summarized his ideas, including that the Earth had originally been formed by solar matter. "On the basis of the Dirac–Gilbert equation," he said, "it can be shown that the earth originated from the sun." Moreover, "The Dirac–Gilbert results permit the establishment of the correct expansion mechanism, and also the computation of the rate of expansion from physical data."[143]

Egyed continued until his death in 1970 to support the expanding Earth and Dirac's $G(t)$ hypothesis, such as is evident from a textbook in geophysics published in 1969. "From Ramsey's model of the Earth and Dirac's cosmology," he wrote, "follow directly the expansion of the Earth."[144] In the book's preface he referred in general terms to his lack of belief in the uniformitarian methodology that still governed geophysical thinking. Since the days of Lyell geologists had taken for granted that the laws of physics were permanent, but according to Egyed this basic assumption of uniformitarianism or so-called actualism was unwarranted:

> Our present physics is based on observations reaching back to about a few centuries, and yet this physics, containing no time parameter, is extended to a time span of a few billion years. The validity of this principle of physical "actualism" has so far never been proved. The author is convinced that the causes of many of the contradictions of the physics of the solid Earth lie in the misunderstanding and abuse of this principle.

Egyed thought to have found additional evidence for Earth expansion not only in paleomagnetism but also in a suggestion made by John Wells, a geology professor at Cornell University and a specialist in fossil corals. Wells' idea was that the number of days per year in the geological past—and hence the speed of rotation of the Earth round its axis—could be inferred from the variation in the deposition of calcium carbonate in fossil corals.[145] By comparing the tiny daily rings with the broader annual bands he found that in the Devonian the year consisted of 400 days, or that in the course of 370 million years the Earth's rotation had slowed down from 22 to 24 hours. Wells' discovery was supplemented and extended by Colin Scrutton, a palaeontologist at the British Museum, who in 1965 suggested that some of the growth-rings were monthly, related to the lunar cycle. Scrutton concluded that in the Middle Devonian the year contained 13 lunar months each of 30.5 days.[146]

---

[142] Ranalli (1971).

[143] Egyed (1961b), p. 427.

[144] Egyed (1969b), p. 279.

[145] Wells (1963).

[146] Scrutton (1965).

Assuming that the faster rotation of the Earth in the past could be ascribed to a smaller radius, Egyed derived from Wells' preliminary coral data that the annual increase of the Earth's radius in the Upper Carboniferous was $\Delta R = 0.58$ mm and $\Delta R = 0.74$ mm in the Middle Devonian.[147] This agreed nicely with his favoured slow expansion rate, but unfortunately the uncertainties in the fossil corals method turned out to be too great to warrant Egyed's conclusion. Although Jordan thought that the expansion rate was much greater, he nonetheless found the evidence from fossil corals to be a "very attractive" argument for the expanding Earth.[148] So did Holmes, according to whom Wells' method placed the subject of the slowing down of the Earth's rotation on a firm basis.[149]

The work of Wells and others on the growth rings of corals as a method of determining the Earth's rotation in the past attracted considerable attention. Wells gave an account of the method at the NASA Earth–Moon conference in 1964, where also Dicke was present, speaking on the Earth's rotation.[150] While Dicke concluded in favour of $G(t)$, another of the participants, the leading British geophysicist Keith Stanley Runcorn, expressed his lack of confidence in the expanding Earth hypothesis. Runcorn argued that Wells' growth rings method in combination with astronomical data ruled out the fast expansion models of Carey and Heezen, which he referred to as "some of the wilder theories of the earth's evolution."[151] The longer days in the past he attributed to tidal friction entirely. Runcorn also considered the slow expansion based on Dirac's $G(t)$ hypothesis to be unlikely.[152] As a result of his own and others' research work in paleomagnetism, Runcorn had in the late 1950s abandoned the fixist view of the Earth and turned into an enthusiastic advocate of continental drift. He agreed with Holmes in praising the wide-ranging significance of Wells' method: "From considering the tiny lines on the skeleton of an animal which lived 370 million years ago we are led to the consideration of profound problems of cosmology, the evolution of the earth, and the formation of the moon."[153]

Bruce Heezen, a marine geologist at Columbia University's Lamont Earth Observatory collaborated with the cartographer Marie Tharp in mapping the Mid-Atlantic Ridge. He was another pioneer of the new theory of the expanding Earth, which he and Tharp for a period preferred over the theory of continental drift.[154] Heezen first discussed the expansion alternative in 1957 but at the time without publishing on the idea. In a talk of 1958 which was published the following year he cautiously advocated an expansion of the mantle, stating that the hypothesis was worthy of serious investigation. The geologist and oceanographer Henry

---

[147] Egyed (1969b), p. 278.

[148] Jordan (1971), p. 115.

[149] Holmes (1965), pp. 972–975.

[150] Wells (1966).

[151] Runcorn (1967), p. 11.

[152] Runcorn (1964). See also Frankel (2012d), pp. 224–232.

[153] Runcorn (1967), p. 11.

[154] See Barton (2002) on the Heezen–Tharp collaboration.

Menard was an active participant in the plate tectonic revolution. "Bruce was talkative about the possibility of an expanding earth, but … he lacked any strong conviction in the period from 1958 to 1960," he recalled. "I kept waiting for him to publish a proper scientific paper exposing his hypothesis to critical review. Meanwhile it was hard to take him seriously."[155]

Over the next several years Heezen supported in a rather sketchy way the idea of a rapid expansion.[156] In a book published on the occasion of the fiftieth anniversary of Wegener's drift hypothesis he spelled out the essence of the expansion picture of the Earth:

> Under this hypothesis, a sialic crust differentiated early in the history of the earth and originally formed an essentially continuous outer shell. After this sialic crust solidified it was broken up by the continued expansion of the interior of the earth. Mantle material, reaching the surface in the cracks between the sialic blocks, formed the simatic ocean floors. As the earth continued to expand, the oceans gradually grew wider, while the sialic continents remained of nearly their original surface area. … This explanation would account for the displacement of the continents one from another without having to assume that the continents had drifted through oceans.[157]

Although Heezen did not come up with a definite figure for the expansion rate, he thought that the rate might be as great as the one proposed by Carey, perhaps amounting to a 45 % increase in the Earth's surface area since the Paleozoic, or what amounts to $dR/dt \cong 7$ mm per year. As to the cause of the expansion, Heezen referred to a combination of the Dirac–Dicke hypothesis of $G(t)$ and Egyed's idea of density changes within the interior of the Earth. "A decrease in the force of gravity combined with internal density changes would produce a very large expansion," he declared.[158] Still in 1960 Egyed was unaware of his fellow expansionist at Columbia University. Under Tharp's influence Heezen eventually abandoned the idea of an expanding Earth for a form of continental drift in the late 1960s.

3.5 Sympathizers of Expansionism

Arthur Holmes found Egyed's ideas to be interesting and promising, including his use of Dirac's cosmological theory. In one of his letters to the Hungarian geophysicist Holmes referred to the steady-state theory of the universe which at the time was much discussed in Britain as a possible alternative to the evolution theories based on the field equations of general relativity. The theory aroused a great deal of attention and was often considered controversial, or even provocative. According to Holmes[159]:

---

[155] Menard (1986), p. 149.

[156] Frankel (2012b), pp. 393–427.

[157] Heezen (1962), p. 283.

[158] Heezen (1960), p. 110.

[159] Letter of 30 August 1959, as quoted in Frankel (2012b), p. 287. On the reception of the steady-state theory in Great Britain and elsewhere, see Kragh (1996).

If the "steady state" hypothesis of the Universe turned out to be correct (which heaven forbid!) surely G would remain constant? However, apart from that very doubtful possibility, it seems reasonably certain that G is decreasing with time. It may even turn out that all the other evidence for an expanding earth is also evidence that G is decreasing and that the "steady state" concept is wrong. But that is probably looking too far ahead. Meanwhile a varying G provides much that we need, though I wonder if it would be enough, by itself.

Recall that in his youth Holmes had been in favour of a classical cyclic or steady-state conception of the universe (Sect. 1.2). He now expressed his strong dislike of the steady-state theory proposed by Fred Hoyle and others, even indicating that if evidence from the Earth proved G to vary, this theory might be refuted.

In the revised 1965 edition of his influential textbook *Principles of Physical Geology* Holmes dealt in some detail with the expanding Earth and its relation to the contemporary cosmological debate.[160] Writing at a time when the cosmic microwave background had not yet entered the picture—his book was prefaced October 1964—he referred to the "Big Bang theory" in the meaning of "one supreme act of Creation followed by an explosive expansion of the matter and energy then created." Holmes may have been the first geophysicist using the term "big bang," which was originally coined by Hoyle in 1948 but sparingly used until the late 1960s. Evidently interested in cosmology, Holmes referred to books and articles by several cosmologists, including H. Bondi, P. Dirac, G. McVittie, and C. Gilbert. The name of Jordan did not appear in his book.

Emotionally Holmes preferred an eternally cyclic or pulsating universe, which he believed "makes an aesthetic appeal to many minds." One of the minds was obviously Holmes'. While not embracing the big-bang theory, Holmes did not embrace the steady-state theory either. In line with what he had written to Egyed, he thought that the expanding Earth contradicted the cosmology advocated by Bondi, Gold, Hoyle and a few others:

> Fortunately our own concern, as geologists, is with the Earth, and our geological interest in the steady state hypothesis lies in its implication that G, the constant of gravitation, also remains steady and really *is* a constant. But, if so, the expansion of the earth is left without an explanation. Conversely, and this is of particular importance to astronomers, the expansion of the earth is a powerful argument against the steady state.

Holmes was not the only geologist who referred to cosmology and had a preference for a cyclic universe. In a textbook of 1976 the geophysicist Adrian Scheidegger, professor at the Technical University in Vienna and a former student of J. Tuzo Wilson, included a section in which he described the new big-bang scenario. He commented:

> Of course, we do not know what the Universe might have been like before the original "big bang." The latter might be part of a "pulsation" of the Universe which might alternate between phases of expansion and contraction. Thus, a "stationary" state might perhaps be

---

[160] Holmes (1965), pp. 983–987.

present after all, in which one "big bang" would follow the next in intervals of about 20×10^9 years.[161]

Although the steady-state cosmological theory was not based on the field equations of general relativity, a varying gravitational constant was indeed incompatible with it. But Holmes was wrong nevertheless, for a varying G was also inconsistent with cosmological models based on the Friedmann equations of general relativity, whether of the big-bang kind or of the cyclic kind. In other words, $G(t)$ could not be used to discriminate between steady-state cosmology and cosmological models based on Einstein's field equations. Varying gravity was simply not an issue in the extended cosmological controversy between the two world systems. Much like Egyed, Holmes uncritically accepted Gilbert's claim that Dirac's $G(t)$ hypothesis was consistent with the standard theory of general relativity.[162]

Holmes was at the time sympathetic to the expanding Earth, if by no means committed to the hypothesis. He advocated a slow expansion of the kind Egyed had proposed but rejected the rapid expansion argued by Carey and Heezen. His preferred value for the increase in radius was only 0.5 mm per year or 100 km in the course of the past 200 million years, whereas Carey's value for the same period was 8 mm per year. It is possible that Holmes came independently to the idea of Earth expansion which he may have inferred from studies of the first erosion processes some 3 billion years ago.[163]

According to Holmes, the major role of the expansion was not to move the continents, but to provide energy for the mantle convection which, in his view, was complementary to global expansion. In broad agreement with Egyed he suggested that there was no reason to choose between continental drift and the expanding Earth: "Convection does not exclude global expansion. Global expansion does not exclude convection. And the combination is stronger than either separately."[164] Not only could convection currents be combined with the expanding Earth, as in the theories of Dicke and Holmes, they also appeared in some of the attempts to explain the origin of oceanic ridges on the basis of the traditional picture of a nearly permanent Earth.

At the end of his book, Holmes referred to the possibility of a cosmological cause for the terrestrial expansion brought about by a relief of pressure:

> There seems to be only one possibility in the light of present knowledge: that the terrestrial force of gravity, which can be represented by g, has systematically decreased as the earth has grown older. This variation of g with time could be brought about in either of two ways. The universal constant of gravitation, G, may have decreased with time, as inferred by P. A.

---

[161] Scheidegger (1976), p. 103. Oscillating models of the kind described by Scheidegger were discussed by several astronomers and cosmologists but were generally seen as somewhat speculative as many-cycle models could not be justified by the equations of general relativity. Moreover, they presupposed space to be closed, which lacked observational evidence.

[162] So did a few other geologists, see for example Stewart (1970), p. 413.

[163] Egyed (1961b), p. 432, referred to a personal communication from Holmes and stated that Holmes had in this way estimated an expansion rate of 0.4 mm per year.

[164] Holmes (1965), p. 967.

M. Dirac in 1938; or matter may have steadily vanished from every part of the earth (and from all other material throughout the universe), as proposed by R. O. Kapp in 1960.[165]

The reference at the end of the quotation was to Reginald Otto Kapp, a British emeritus professor of electrical engineering, who in 1960 published a speculative and semi-philosophical cosmological theory which included continual disappearance of matter as well as creation of new matter. His theory went back to a paper of 1940 and in its later version Kapp suggested that it was in the same tradition as the steady-state theory of the universe.[166] He also derived from his theory various geophysical consequences, including that the Earth was *contracting* from a much larger Earth in the past. According to Kapp, his theory explained the formation of mountain ranges and it "transforms the Wegener Hypothesis from the category of ad hoc hypothesis into that of inference."[167] Although Holmes did not accept Kapp's amateurish cosmology—and presumably even less his amateurish geophysics—it is a little surprising that he took it seriously and dealt with it on par with Dirac's cosmology.

Not all expansionists found it necessary to suggest a cause, either cosmological or in terms of radioactivity or geochemistry, for the expansion of the Earth. According to the Canadian geophysicist Alan Beck, the energy needed for the expansion could easily be provided by the Earth itself. Assuming the Earth to be of uniform density and constant mass, its gravitational potential energy is

$$E_{grav} = -\frac{3}{5}\frac{GM^2}{R}.$$

From dE_{grav}/dt it is seen that an increase in R requires an injection of energy. As Beck showed, the situation is different for a non-uniform Earth, which can expand and loose gravitational energy at the same time. Using a density-radius contour of the quadratic form

$$\rho(r) = \rho_0\left(1 - kr^2\right), \quad 0 < r < R,$$

he found the potential energy of the Earth to be -2.5×10^{39} ergs. Beck argued that "expansions of the order of 100 km seem quite possible without postulating any source of energy."[168] On the other hand, he found a radial increase of the order of 1000 km to be improbable even if the effects of radioactivity were taken into account. Such an expansion would need a non-conventional cause of some kind, perhaps in the form of varying gravity. Without taking the possibility of a decreasing G into account, two physicists from the University of Utah, Melvin Cook and

---

[165] Holmes (1965), p. 983.

[166] See Wesson (1973), pp. 28–29 and Kragh (1996), pp. 151, 196–197. Also Stewart (1970) dealt with Kapp's speculative cosmology.

[167] Kapp (1960), p. 243.

[168] Beck (1960, 1969).

A. Eardley, independently made calculations of the energy required for an expanding Earth. Their result, that neither radiogenic heat nor phase changes were able to cause an expansion as substantial as the one suggested by Carey, agreed with the conclusion of Beck.[169]

Another Canadian geophysicist, John Tuzo Wilson at the University of Toronto, had a high reputation in the community of earth scientists. Originally studying mathematics and physics he switched to geology and in 1930 he graduated as the first Canadian with a degree in geophysics. After a stay at Cambridge University Wilson went to Princeton, from where he obtained his Ph.D. in 1936. While at Princeton he became friends with Harry Hess. Since the late 1940s Wilson had defended the contractionist view of the Earth and attacked the mobilist view of drifting continents which, he argued, was empirically inadequate, lacked a physical mechanism, and contravened the methodological desideratum of uniformitarianism. As late as 1959 he confirmed his stance against mobilism, only to change one or two years later to become a champion of sea floor spreading, mantle convection and continental drift.[170] Wilson then quickly emerged as one of the leaders of the new theory of plate tectonics. The remarkable shift from one paradigm to another was facilitated by a brief flirtation with the idea that the Earth had expanded as a result of decreasing gravity.

In a paper in *Nature* of March 1960, Wilson cautiously supported the slow version of the expanding Earth hypothesis and its explanation in terms of a decreasing gravitational constant. He suggested that expansion might be the cause of the widening of the ocean basins and explain the formation of ridges. Citing Dirac, Teller, Dicke and Egyed (but not Jordan) he admitted that a slow expansion might be caused by other factors, such as phase changes in the interior of the Earth. On the other hand, "a decrease in G remains an inviting idea."[171] Wilson was well aware of Dicke's argument connecting varying gravity and the expanding Earth from the magazine *American Scientist*: an article that Dicke published on this subject in 1959 was preceded by an article by Wilson based on the contracting Earth.[172]

On the basis of recent estimates of the area covered by the mid-oceanic ridges Wilson calculated that their formation would require an increase of the Earth's circumference of about 6 % in "all geological time," which he found to be "close to

---

[169] Cook and Eardley (1961).

[170] See details in Frankel (2012d), pp. 3–50. According to "Planck's principle" as used by Thomas Kuhn and some sociologists of science, radical scientific change—from one paradigm to another—is a non-rational event reserved for scientists of a young age. Older scientists are supposed to stick to the established paradigm until they pass away. The principle has many exceptions, though, one of them being J. Tuzo Wilson, who was 53 years old when he turned to continental drift. Another example is Carey, who at the age of 45 changed to the expanding Earth after having defended continental drift for two decades. On Planck's principle, a name that derives from the autobiography of Max Planck, see Blackmore (1978).

[171] Wilson (1960), p. 882.

[172] Dicke (1959a), Wilson (1959), Menard (1986), p. 173.

Dicke's estimate." If "all geological time" is taken to be 4 billion years, the annual increase in radius would be about 0.8 mm. Wilson criticized the theory of continental drift and also Heezen's view of rapid expansion, which received support from only a few scientists. In his paper of 1960 Wilson expressed serious interest in the expanding Earth and $G(t)$, which he found could explain global tectonics as well or even better than the contracting Earth. He explicitly rejected expansion as an explanation for continental drift, should it be real. However, Wilson's interest in the expanding Earth hypothesis turned out to be nothing but a brief flirt. By the fall of 1961 he had convinced himself that the continents were moving apart in accordance with Wegener's old idea. He did not return to speculations about gravity varying in time or to the expanding Earth. These topics simply did not appear in a comprehensive popular account of the new theory of continental drift that he wrote for *Scientific American*.[173]

On the other hand, Wilson seems to have recognized the expanding Earth model as a serious alternative to other theories of the Earth. In a paper of 1963 on the origin of the Hawaiian Islands he distinguished between four and not only two major whole-Earth theories. His typology was this:

1. The traditional view of a rigid Earth supported by Jeffreys and "most geologists in the northern hemisphere."
2. The expanding Earth supported by Dicke, Heezen and Carey in different versions.
3. Wegener's theory of continental drift as revived by Runcorn, Blackett and others.
4. The theory of convection cells in the mantle of the Earth, which "provides a new and better mechanism for continental drift than that proposed by Wegener."[174]

Wilson took his study of the Hawaiian Islands to be strong evidence for the last of the four theories.

3.6   Discussions Pro et contra

The varying-G hypothesis became associated with the expanding Earth in the 1950s, largely as a result of the work of Jordan and Egyed. But the first contact between Dirac's hypothesis and the state of the Earth was established in 1948, when Teller argued that the hypothesis led to a temperature in the past that was too high to agree with paleontological evidence. Even if one accepted Teller's argument and concluded that Dirac's $G(t)$ hypothesis was wrong, it did not follow that the size of the Earth had not increased over geological time. The climatological argument had little effect on the viability of the expanding Earth hypothesis.

---

[173] Wilson (1963a).

[174] Wilson (1963b), p. 864.

In the period around 1960s there was no consensus with regard to the temperature of the Earth in the past. The traditional view going back to the French naturalist Georges Leclerc (Comte du Buffon) in the mid-eighteenth century was that the Earth had been formed in a molten state and gradually cooled. However, there were also scientists who questioned if the climate of the ancient Earth had really been warmer than today. According to the authoritative *Handbuch der Physik* (Handbook of Physics) "it is safe to assume that the surface temperature of the Earth has remained practically constant throughout the whole of geologic time."[175] The temperature might even have increased. Referring to astrophysical theory and mineralogical evidence some geologists suggested that the Precambrian had been characterized by a very *low* surface temperature which since then had gradually increased to the present average value of 14 °C. This view was advocated by Alfred Ringwood, a young Australian geochemist and cosmogonist, and subsequently defended in a modified version by Fairbridge.[176]

In his classic *Structure and Evolution of the Stars* Martin Schwarzschild had shown that the Sun's luminosity was slowly increasing and had initially, for some 5 billion years ago, been 1.6 times smaller than its present value. His result assumed $G =$ constant. Without suggesting an answer, Schwarzschild asked: "Can this change in the brightness of the sun have had some geophysical or geological consequences that might be detectable?"[177] Öpik and Hoyle arrived independently at the conclusion that the energy output of the Sun had increased over geological time.[178] Öpik estimated that shortly after the formation of the Earth the Sun's luminosity was 85 % of the present value and the surface temperature of the Earth about -17 °C; 1.5 billion years later the values had increased to 92 % and $+5$ °C, respectively. Although Schwarzschild was aware of the $G(t)$ hypothesis he did not refer to it in either his 1956 work with Howard and Härm or in his book of 1958. Only in 1964, in his work with Pochoda, did he take up the subject.[179] Öpik too waited until the mid-1960s to comment on the Dirac–Jordan hypothesis.

In May 1957 Hoyle participated in a conference on stellar populations in the Vatican Observatory, Castel Gandolfo, Rome. Among the attendants were also Martin Schwarzschild, Otto Heckmann, Walter Baade, Georges Lemaître and Bengt Strömgren. Hoyle used the occasion to present computations of the Sun's evolutionary track, plotting the solar luminosity and radius as functions of time. He found the initial temperature of the Sun to be fainter than at present, "in agreement with the recent estimate of Schwarzschild, Howard and Härm." Although Hoyle did not relate the increased solar luminosity to the Earth's paleoclimate, and also did not refer to Teller's scenario based on decreasing gravity, he did comment on the

[175] *Handbuch der Physik*, vol. 47 (Berlin: Springer-Verlag, 1956), p. 390

[176] Ringwood (1961), Fairbridge (1964, 1966). On Ringwood's work, see Brush (1996c), pp. 147–149.

[177] Schwarzschild (1958), p. 207.

[178] Öpik (1958, 1965), Hoyle (1958).

[179] Pochoda and Schwarzschild (1964). See Sect. 2.5.

Fig. 3.5 Fairbridge's comparison of Dicke's curve for the early Earth temperatures and the contrasting results based on constant gravity. *Source:* Fairbridge (1966), p. 146. Reproduced with the permission of John Wiley and Sons

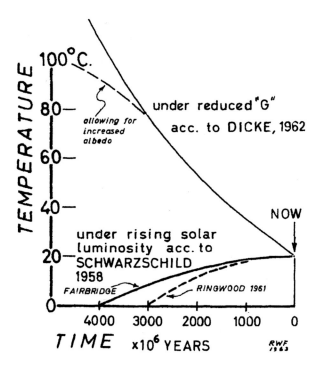

climate of the Earth in the far future. Hoyle's contribution to what later would be called "physical eschatology" was this: "The future span available to life on Earth can be seen to lie in the region 2–3 billion years—after a further 5 billion years the oceans will boil and no life will surely persist then."[180]

Picking up on Schwarzschild's question, Fairbridge argued that 4 billion years ago the mean temperature of the Earth's surface was about 0 °C (Fig. 3.5). This was in fact what Schwarzschild and his Princeton collaborators Robert Howard and Richard Härm had earlier suggested from their model calculations of the Sun:

> In the early pre-Cambrian era, two billion years ago, the solar luminosity was about 20 per cent less than now. The average temperature on the earth's surface must then have been just about at the freezing point of water, if we assume that it changes proportionally to the fourth root of solar luminosity. Would such a low average temperature have been too cool for the algae known to have lived at that time?[181]

As Fairbridge pointed out, this climate scenario was diametrically opposite to the one proposed by Teller, Dicke and others on the basis of a decreasing

---

[180] Hoyle (1958), p. 230. See also Hoyle (1994), pp. 299–303 for the Rome conference. On so-called physical eschatology, see Kragh (2011), pp. 325–354.

[181] Schwarzschild et al. (1958), p. 241.

gravitational constant. Fairbridge concluded: "The evidence of paleoclimatology favors the concept of expansion, but is opposed to change of the gravitational constant."[182]

The latter part of Fairbridge's conclusion was substantiated by two German astronomers at the University of Bochum, Walter Eichendorf and Michael Reinhardt. In the 1970s geochemically based estimates of paleotemperatures indicated that the surface temperature of the Earth 3.4 billion years ago had been between 65 and 95 °C, which was about the value to be expected from Teller's argument already 2 billion years ago. Eichendorf and Reinhardt used the data to propose that the relative change in G was less than 2×10^{-11} per year. Moreover, they elaborated on Teller's brief remark of 1948 that, if mass creation according to $M \sim t^2$ were assumed it would have been too cold on the Earth. Instead of diminishing as $T \sim t^{-9/4}$ the Earth's surface temperature would increase according to

$$T \sim t^{1/4}.$$

As a result, since 3.4 billion years ago the temperature would have been about or below the freezing point of water. The two German astronomers recommended that "Dirac's large numbers hypothesis should be buried with appropriate honours."[183] Perhaps it should, but this is not what happened.

The early work of Schwarzschild and his collaborators can be seen as an anticipation of what came to be known as the "faint young Sun paradox," or what today is often abbreviated the FYS-paradox. This paradox or problem is the apparent contradiction between strong evidence for liquid surface water on the Earth more than 3 billion years ago and the prediction from reliable solar evolution models that the energy input from the Sun was at the time 25–30 % lower than today. Assuming an unchanged atmospheric composition the result of a smaller solar luminosity will be a frozen Earth during the first 2 billion years of the existence of our planet. And yet it is known that in the Archean there were oceans of liquid water. The solar models are considered robust and very reliable, implying that there must have been one or more effects in the past compensating for the faint young Sun. What these effects were is still a matter of some debate.

The problem was first highlighted by the Cornell astronomers Carl Sagan and George Mullen in 1972 and subsequently generated a large number of scientific papers.[184] While Sagan and Mullen thought that the compensating agent was atmospheric ammonia, a powerful greenhouse gas, other specialists disagreed. To solve the "glaring conflict between solar models and the biological and isotopic

---

[182] Fairbridge (1964), p. 83, who seems to have been unaware of the Schwarzschild–Howard–Härm paper.

[183] Eichendorf and Reinhardt (1977), p. 537. Eichendorf and Reinhardt did not refer to either Fairbridge or Schwarzschild.

[184] Sagan and Mullen (1972). For a detailed review of the faint young Sun problem, see Feulner (2012).

history of the earth" Michael Newman and Robert Rood considered the possibility that G varied in accordance with the Brans–Dicke theory, but as slowly as $dG/Gdt = -3 \times 10^{-12}$ year^{-1} corresponding to an omega parameter as large as $\omega = 12$.[185] However, they found the possibility of an explanation based on varying gravity to be unlikely and instead argued for a stronger greenhouse effect in the past. Today most experts agree that cosmological effects are of no relevance to the faint young Earth problem, and yet a few scientists have advocated varying gravity as a possible solution.[186] Whatever the precise answer to the problem, the many analyses demonstrate how woefully inadequate and oversimplified Teller's old argument was.

While considerations of paleoclimatology were not directly related to the expanding Earth, the situation was different with regard to paleomagnetic research. Although the results derived from measurements of paleomagnetism were far from unambiguous, the measurements played an important role in the debate concerning the expansion hypothesis and its relation to continental drift. Whereas Carey thought that his idea of rapid expansion received support from paleomagnetism, most experts disagreed. Allan Cox and Richard Doell at the U.S. Geological Survey were in the early 1960s cautiously moving toward support of continental drift. They could find no paleomagnetic evidence for an expanding Earth, although they obtained a radius for the Earth in the Permian (6310 km) that was slightly less than the present radius. The result did neither confirm nor disconfirm Egyed's slow expansion rate, but it ruled out the much greater rate argued by Carey.[187]

A similar conclusion was reached by Edward Irving, a British-Canadian geologist and authority in paleomagnetism. In his monograph *Paleomagnetism* of 1964 he discussed Egyed's and Carey's arguments for the expanding Earth, concluding that "The inconsistency with Carey's hypothesis is substantial and appears therefore to invalidate Earth expansion as a cause of continental drift."[188] Although most researchers in paleomagnetism agreed with the verdict of Cox, Doell and Irving, not all did. Carey countered that Cox and Doell had misunderstood his model.[189] Had they applied their test to the correct model, they would have arrived at a Permian radius of about 4500 km.

Among the few earth scientists in favour of rapid expansion was, apart from Carey and Heezen, the Dutch geophysicist D. van Hilten, who based his support on evidence from paleomagnetism. In qualitative agreement with Carey he derived the values for the Earth's radius in the past which are tabulated in Table 3.2. Also in

---

[185] Newman and Rood (1977). For the Dicke–Brans ω parameter, see Sect. 2.7.

[186] See Tomaschitz (2005), according to whom "The age of the universe has a substantial imprint on planetary paleoclimatology."

[187] Cox and Doell (1961).

[188] Irving (1964), p. 292.

[189] Carey (1961, 1976, p. 185).

Table 3.2 Van Hilten's
rapidly expanding Earth

Period	% of present radius	R (km)
Carboniferous	79.5	5065
Permian	83.0	5288
Triassic	87.0	5543
Jurassic	89.5	5702

Data from van Hilten (1963, 1964)

agreement with Carey, he assumed that during the Earth's expansion the area of the continents and the continental shelves remained unaltered, whereas the oceans grew bigger. Although van Hilten's conclusion was broadly criticized, it received the support of Holmes and Jordan.[190]

Contemporaneously with van Hilten the Australian geophysicist Martin Ward used a generalization of a magnetic method developed by Egyed to estimate the paleoradius of the Earth R relative to its present value R_0. He found R/R_0 to have been 1.12 in the Devonian, 0.94 in the Permian, and 0.99 in the Triassic. Taking the uncertainties into regard, Ward concluded that there had been no significant change in R since the Permian.[191] While his result made Carey's rapid expansion "unlikely," the accuracy of the method was not good enough to rule out the slow expansion proposed by Egyed. The conclusions of van Hilten and Ward were quite different and yet they were based on largely the same methods, data and assumptions. According to van Hilten, Ward's interpretation of the data was objectionable and his conclusion unreliable.[192] He maintained that the Earth had been smaller in the past.

The disagreement between the two geophysicists was methodological and did not concern the cause of the expansion, if it were real. Neither van Hilten nor Ward referred to the $G(t)$ hypothesis. Van Hilten did not speculate about the cause of the expansion, but believed that it was sufficient to account for the relevant geological data. In a review article in the first issue of the new journal *Tectonophysics* he wrote: "Besides global expansion, no additional mechanisms, as for instance that of convection currents in the mantle, seem to be required to explain the relative movements between the continents."[193] However, his method and arguments were severely criticized by two of his compatriots, the Amsterdam geophysicists Jan Hospers and S. Van Andel, who in agreement with Ward concluded that Carey's large expansion rate was ruled out and that only the slow expansion advocated by

---

[190] Holmes (1965), p. 209 and Jordan (1971), p. 92. See also Frankel (2012c), p. 230.

[191] Ward (1963).

[192] Van Hilten (1965).

[193] Van Hilten (1964, p. 61, 1965). *Tectonophysics*, subtitled *International Journal of Geotectonics and the Geology and Physics of the Interior of the Earth*, was established in 1964 by the Elsevier publishing company with the aim of promoting greater cooperation between geologists and geophysicists. Its editorial board included W. Brian Harland (Cambridge), S. Keith Runcorn (Newcastle) and J. Tuzo Wilson (Toronto).

Egyed, Dicke and Kenneth Creer remained a possibility.[194] In later studies the two Dutchmen confirmed at a 95 % confidence level that any Earth expansion rate exceeding 2.5 mm per year could be rejected.[195] Thus the paleoradius of the Earth could not possibly have attained the values suggested by van Hilten.

In March 1964 a major conference on continental drift sponsored by the Royal Society took place in London. The organizers were Blackett, Runcorn and the leading Cambridge geophysicist Edward Bullard, who all were in favour of the theory of drifting continents. The same was the case with the large majority of the participating earth scientists, while there was almost no support for the theory of an expanding Earth.[196] According to one of the participants, the British geophysicist Dave C. Tozer, theories of Earth expansion were of an ad hoc nature, which "require either rather fantastic physical properties for the major constituents of the planet, or a premature meddling with the foundations of physics." For this reason, he continued, they should be dismissed "until their assumptions can be independently justified or the inadequacy of convection theories demonstrated."[197] He probably had in mind the class of varying-G theories, which did meddle with the foundations of physics.

Bullen belonged to the majority of geophysicists who saw no merit in the idea of either an expanding Earth or a gravitational constant that had been greater in the past. On one occasion he justified his view methodologically by referring to what is known as Occam's razor: "In accordance with the simplicity postulate of scientific inference, it is appropriate to treat entities like G as constant so long as the available evidence does not suggest otherwise."[198] On the other hand, "that is not to say that G is necessarily independent of time." Expansionist scientists increasingly fought an up-hill battle and yet the theory of the expanding Earth with or without the $G(t)$ hypothesis lived on.

Kenneth Creer, a physicist at the University of Newcastle and a former student of Keith Runcorn, became interested in the expanding Earth after having met Carey in the early 1960s. However, contrary to Carey and van Hilten he believed that most of the expansion took place during the Archaeozoic (Archaeon) and that it played very little role in the Mesozoic era when Pangaea broke into two parts, the ancient continents of Laurasia and Gondwanaland. Creer also disagreed with van Hilten's small radius of the Permian Earth and generally thought that the hypotheses of a rapid expansion confined to the more recent geological ages was "most improbable."[199] He assumed that expansion had occurred over a period of 3.5 billion years.

---

[194] Hospers and Van Andel (1967). Van Andel earned his doctoral degree from Amsterdam University in 1968 with a dissertation on "A Test of Earth Expansion Hypotheses by Means of Paleomagnetic Data."

[195] Van Andel and Hospers (1968), Hospers and Van Andel (1970).

[196] See Frankel (2012d), pp. 162–170.

[197] Tozer (1965), p. 253.

[198] Bullen (1975), p. 345.

[199] Creer (1965a, b).

It started at the time when the Earth might have been entirely covered by the continental crust and had a radius of only 0.55 R_0, where $R_0 = 6378$ km is the present value. By simulating the expansion process he estimated that in the early Paleozoic, some 544 million years ago, the Earth would have swelled to 0.95 R_0 and to about 0.97 R_0 by the end of the era. For the average rate of expansion he suggested the value 0.75 mm per year. Contrary to many other expansionists he did not regard expansion as the principal cause of the formation of mountains.

Although Creer did not refer to the Jordan–Dicke–Egyed hypothesis of a decreasing gravitational constant, he did suggest that the expansion of the Earth might have a cosmological origin of some kind:

> For an adequate explanation we may well have to await a satisfactory theory of the origin and development of the universe. In the meantime, we should beware of rejecting the hypothesis of expansion out of hand on grounds that no known sources of energy are adequate. It may be fundamentally wrong to attempt to extrapolate the laws of physics as we know them to-day to times of the order of the age of the Earth, and of the universe.[200]

In February 1965, when Creer's paper appeared in *Nature*, the more satisfactory theory of the origin of the universe was on its way in the form of the revived hot big-bang scenario. But contrary to what Creer suggested, this theory (contrary to the steady-state theory) built on extrapolations of the known laws of physics. In another paper he discussed the $G(t)$ hypothesis as proposed by Dirac and more recently, so he claimed, by Hoyle and Jayant Narlikar.[201] Creer was not a convinced expansionist, but he thought that the hypothesis deserved serious consideration. Much like Dicke he argued that slow expansion could not be the principal cause or source of either continental drift or orogenic processes:

> I think that expansion should be regarded as something which may have been gently, but persistently, occurring in the background. There may be little obvious geological evidence of expansion: most of this could easily have been obscured by more vicious and rapid processes such as continental drift and orogeny.[202]

As to drift, Creer thought that it would soon be satisfactorily explained by convection in the mantle. Dietz, for one, was not impressed by Creer's claim to have fitted the present continents on a small globe covered entirely by a sialic crust. Not only was the claim based on false assumptions, so Dietz objected, it also implied that nearly the entire hydrosphere was of recent origin. For this and other reasons he considered Creer's version of the expanding Earth to be no less untenable than Egyed's version. He wrote it off as "an absurdity."[203]

Contrary to Dietz, the London geologist Raymond Dearnley largely agreed with Creer's picture. Comparing various methods of estimating the growth of the Earth, he argued that the radius of the Earth had increased from 4400 km to its present

---

[200] Creer (1965a), p. 539.

[201] Creer (1965b), who seems to have misunderstood the Hoyle–Narlikar theory, which did not, in fact, operate with a varying G. On Hoyle's later ideas, see Sect. 4.1.

[202] Creer (1965a), p. 543. For Dicke's view see Dicke (1961b, 1962a) as discussed in Sect. 2.7.

[203] Dietz (1967), p. 235. For Dietz's dismissal of Egyed's theory, see Sect. 3.4.

value R_0 over a course of 2.75 billion years; 600 million years ago the radius was about 6000 km.[204] Dearnley concluded that the methods "strongly suggest a relatively uniform rate of expansion of the Earth's radius of about 0.65 ± 0.25 mm per year as far back as 4500 million years."[205] At the 1967 Newcastle meeting on geophysics he repeated the conclusion, noting that the value was almost the same as the cosmological Hubble expansion rate.[206] Without committing himself, Dearnley was favourably inclined to a $G(t)$ explanation based on the theories of Dirac, Gilbert and Dicke. However, like several other protagonists of the expanding Earth he pointed out that a decrease in G alone would be insufficient to account for the fast expansion. After all, the density of solid matter is primarily determined by the electrostatic force 10^{39} times stronger than the gravitational force. A weaker G in the past would have to be supplemented by some other cause, most likely phase changes in the Earth's mantle such as discussed by Egyed, Creer and several other geophysicists and geochemists.

Although Dearnley must have known about Jordan's work, he failed to mention the German physicist. This was not unusual in the English and American literature. On the other hand, the Canadian geologist Johann Steiner gave full credit to Jordan in an ambitious and somewhat speculative attempt to apply the "Dirac–Jordan effect" to both the Earth and the Milky Way system. "The assumption," he wrote, "that the gravitational 'constant' is universal and constant in time ... should be considered untenable today, and deserves re-examination in the light of geological phenomena."[207] However, Steiner's version of the $G(t)$ hypothesis was unusual in several respects, not least because of his belief that the gravitational constant varied cyclically with a period of about 250 million years. Since he also believed that the size of the Earth depended on the value of G, it followed that the Earth alternately expanded and contracted.

Yet another scientist who discussed the geophysical consequences of Dirac's $G(t)$ hypothesis was Francis Birch of Harvard University, according to whom the radius of the Earth, assuming its mass to be constant, could at most have increased 100 km. If the gravitational constant had decreased from $2G$ to G the radius would have increased 370 km, but such a rate he found unrealistic.[208] There was no agreement with regard to the size and rate of the expansion among the minority of scientists dealing with the expanding Earth. As shown by Table 4.2, expansion rates varied over a large range, from about 8 to 0.03 mm per year.

Adopting Ramsey's phase change hypothesis for the interior of the Earth and Dirac's original $G(t)$ hypothesis, the Japanese physicist Shin Yabushita, at

---

[204] Dearnley (1965).

[205] Dearnley (1966), p. 32.

[206] Dearnley (1969). See also MacDougall et al. (1963).

[207] Steiner (1967), p. 99, who referred to a "personal communication" from Jordan. Ten years later, Steiner came out in support of the expanding Earth (see Sect. 4.4).

[208] Birch (1968).

Kyoto University, computed that changes in radius and gravitational constant were related as

$$\frac{dR}{R} \cong -0.3 \frac{dG}{G}.$$

It followed that at the time when the Earth was formed, G was about 1.5 times greater than at present and R some 700 km less than it present value.[209] The primitive Earth would thus have had $R \cong 5670$ km.

One of the problems originally motivating the idea of Earth expansion was the difficulty of making the moving continents fit together on an Earth of the present size. After continental drift and sea floor spreading had been widely accepted the problem was reconsidered by Robert Meservey, a condensed matter physicist and amateur geophysicist at the Massachusetts Institute of Technology. Meservey argued that the motion of the continents according to global plate tectonics was not topologically possible on a present-sized Earth. On the other hand, the "paradox" might be resolved if "a large expansion of the earth's interior has taken place in the last 150 million years." As to the cause of the expansion he mentioned the possibility of a decreasing gravitational constant, vaguely suggesting that although it or some similar mechanism was "highly conjectural," yet it "cannot be excluded on the basis of present physical knowledge."[210]

It should be noted that the assumption of a varying gravitational constant did not necessarily result in a steadily expanding Earth. The Portuguese geologist F. Machado instead used geological and oceanographic data since the Devonian to suggest that the size of the Earth had pulsated through a series of expansions and contractions.[211] Assuming a period of approximately 200 million years, he wrote the variation in the Earth's radius as

$$\frac{dR}{R} = 0.03 \sin\left(2\pi \frac{t+50}{200}\right),$$

where t is the time measured in millions of years. From this Machado inferred that G had varied in a similar but opposite manner, namely given by

$$\frac{dG}{G} \cong -20 \frac{dR}{R}.$$

Machado wisely avoided commenting on the physical and astronomical consequences of the unlikely pulsation hypothesis. Unlikely it was and yet Machado's suggestion was independently supported by Steiner's study from the same year. The

---

[209] Yabushita (1984).

[210] Meservey (1969), with references to Dirac, Carey, Egyed, Heezen and Wilson. As usual in the English-language literature, Meservey did not refer to Jordan.

[211] Machado (1967).

idea of a pulsating Earth with radial changes related to a new conception of gravity was not new. Based on a speculative "exponential law of gravitation" a theory of this kind was suggested by Anatol Schneiderov at the George Washington University as early as 1943, and similar speculations were forwarded by some Russian scientists.[212] Contrary to the hypothesis of the expanding Earth, pulsating-Earth and pulsating-gravity hypotheses were never taken seriously.

Although Dicke, who was an advocate of the cyclic or pulsating universe, did not refer to a pulsating Earth in his writings, he may have considered the possibility in private. According to Fairbridge, "Dicke (personal communication) remarked that such expansion [of the Earth] might well be pulsatory."[213] The general idea of cycles or pulsations in the history of the Earth goes far back in time but in the more limited sense that terrestrial phenomena such as mountain formation or marine regression were assumed to vary cyclically on an Earth of fixed size. Ideas of this kind were common in the late nineteenth and early twentieth centuries when they were suggested in different forms by, for example, Joly, Chamberlin and the American geophysicist David Griggs.[214] None of these suggestions referred to cyclical changes in the radius of the Earth.

Another unconventional and short-lived hypothesis, proposed by an American chemist in 1980, was that the gravitational constant *increases* rather than decreases with time. The deceptibly simple argument was that according to $GM = Rc^2$ and assuming M to be constant, it follows that $G \sim R$. Since the universe expands, the gravitational constant must increase, for example as

$$G \sim t$$

At the big bang G would be infinitesimally smaller than the present value and thus allow "the instantaneous flying apart of neutrons in the original spherical configuration" of the universe.[215] As pointed out by T. L. Chow, a physicist at Humboldt State University in California, the $G \sim t$ hypothesis had much earlier been suggested by Milne, although in a version with no terrestrial consequences. Chow objected that the proportionality hypothesis would lead to a surface temperature of the Earth varying as

[212] Schneiderov (1943), Carey (1988), p. 145. In Schneiderov's theory the gravitational constant G might be a variable quantity but in a completely different way than in Dirac's $G(t)$ theory. Schneiderov's ideas about gravitation were pre-Einstein and had no impact at all on later developments in either cosmology or geophysics. For pulsating-Earth ideas in Soviet Russia, see Sect. 4.4.

[213] Fairbridge (1964), p. 60.

[214] For geological pulsation hypotheses, see Oldroyd (1996), pp. 182–188.

[215] Levitt (1980), p. 24. The picture of the big bang as an explosion of an original mass consisting of neutrons was part of Gamow's theory in the late 1940s, but in 1980 it was obsolete.

$$T \sim G^{9/4} \sim t^{9/4}$$

instead of the Dirac–Teller variation $T \sim t^{-9/4}$. The primitive Earth would thus have been very cold. Moreover, the $G \sim t$ hypothesis implied a slow contraction of the increasingly warmer Earth. According to Chow, this was contrary to "paleogeographical and other evidence [which] points to expansion and not contraction of the earth."[216] Nothing more was heard of the increasing-G hypothesis.

The British theoretical physicist Paul Wesson wrote in the 1970s several comprehensive reviews of geophysical theories and their relations to cosmology. Contrary to most other physicists and astronomers involved in the dynamics of the Earth, Wesson had an extensive knowledge of the geological and geophysical literature and published some of his papers in journals devoted to the earth sciences. All the same, earth scientists rarely referred to him. According to Wesson, the new plate tectonics was inadequate to account for geophysical data and possibly inferior to the expansion hypothesis.[217] Listing a large number of shortcomings of the global theory of plate tectonics he concluded that "the continents have almost certainly not moved with respect to each other."[218]

In a comprehensive review article of 1973 on what he called "geophysics on a global scale" Wesson came out in support of the expanding Earth, if not of expansion driven by a decreasing G. Perhaps, he wrote, "sea-floor spreading may be something of an illusion caused by the continents sitting still while the globe expands beneath them."[219] Wesson thought that the Earth had expanded from a state with half the present radius, or that its circumference had increased at a rate of about 10 cm per year. None of the existing $G(t)$ hypotheses could explain an expansion of this scale. As an alternative he suggested that the expansion might be caused by continuous creation of matter proportional to the mass of the Earth. What kind of matter creation? Obviously the one provided by the classical steady-state theory was irrelevant, since it only amounted to 10^{-43} g s^{-1} cm^{-3}. Wesson needed matter creation of a rate about 7×10^{-18} g s^{-1} cm^{-3}, and unfortunately there was no experimental evidence whatever for such drastic creation processes.

---

[216] Chow (1981), p. 120. An earlier proposal of a steady contraction of the Earth, but without basing it on an increasing gravitational constant, can be found in Kapp (1960). The notion of a contracting Earth was also defended by R. Lyttleton, see Sect. 4.1.

[217] Wesson (1970).

[218] Wesson (1972), p. 185.

[219] Wesson (1973), p. 43.

Chapter 4
After Plate Tectonics

Jordan and Dicke were not the only cosmologists who thought that varying gravity and other exotic ideas from fundamental physics might be relevant for the earth sciences. In the 1970s Fred Hoyle developed a revised steady-state model of the universe with implications for the history and structure of the Earth. In the same decade Dirac returned to his favourite hypothesis of a decreasing gravitational constant. Attempts to test the $G(t)$ hypothesis in one of its several versions came from physics, astronomy and geology until it gradually became clear that the constant is indeed constant—as far as measurements can tell. In the same period the expanding Earth hypothesis ran out of power and separated increasingly from mainstream geophysics. The hypothesis of a smaller Earth in the past continued to be defended but without being taken seriously any longer by the majority of earth scientists.

Apart from dealing with the declining phase of the two heterodox hypotheses this chapter also compares the two revolutions in science that occurred in the 1960s, namely, plate tectonics and big-bang cosmology. Was it just one of history's many coincidences that the two fundamental theories won acceptance in the same period of time?

4.1 Steady-State Cosmology and the Earth

As we have seen, in the 1960s and 1970s there were in some quarters of the earth sciences a feeling that a proper understanding of the Earth must involve considerations of a cosmological nature. The common denominator of geophysics and cosmology was usually regarded to be the force of gravitation.

Creer, a mainstream geologist, found the $G(t)$ hypothesis to be interesting, but in his view a larger G in the past could not alone have caused a substantially smaller Earth. "Are we, in extending the range of our experience to phenomena occurring during aeons of time, about to be faced with another revolution in physics?" he

© Springer International Publishing Switzerland 2016
H. Kragh, *Varying Gravity*, Science Networks. Historical Studies 54,
DOI 10.1007/978-3-319-24379-5_4

asked. "Perhaps the constants of physics change in time in such a way that the beginning of time, defined as the time of the creation of the universe, is meaningless." Creer did not explain precisely what he had in mind, but perhaps he was thinking of the infinitely large gravitation constant that in a formal sense appeared in Dirac's theory at the singularity corresponding to $t = 0$. According to Creer, it was natural to infer from the $G(t)$ hypothesis that at least some of the other fundamental constants must also have changed in time. His argument was as follows: "Otherwise the sun would have been tens of times brighter when the earth's radius was about half its present value. Thus the earth's surface would have been molten at a time when we have reason to believe that it was solid."[1]

By the mid-1960s the expansion of the universe had been known for than three decades and was regarded as an established fact by the large majority of astronomers. Could the hypothetical expansion of the Earth somehow be related to the factual cosmological expansion? There is no reason to assume that the early advocates of the expanding Earth were inspired by some kind of universe-Earth analogy or related the terrestrial expansion to the one discovered by the astronomers. But in a few cases they compared the two phenomena. It was quite natural for Holmes to include in his textbook of physical geology a section with the title "The Expanding Earth and the Expanding Universe." After all, "New ideas of atomic structure at one end of the scale of dimensions and of the expanding universe at the other, necessarily demand new ideas about the earth herself."[2] At about the same time Fairbridge vaguely suggested a connection: "According to Einstein, the Universe is expanding, and Dirac (1938) concluded that gravitation must decrease with time. ... A test of such a theory would be the demonstration of a slow expansion of our globe."[3]

Carey was another prominent geologist who toyed with the idea of some deep but unspecified connection between the expanding Earth and the expanding cosmos. "This universal dissipation [of continents and oceans] implies an expanding earth," he said, "just as the universal red shift of stellar spectra is taken to mean an expanding Universe."[4] In 1963 a group of four Canadian physicists called attention to the "remarkably close agreement between the rate of increase of the Earth radius and that of the universe according to Hubble's law."[5] According to the empirical

---

[1] Both citations are from Creer (1965b), p. 39. Creer speculated that the permittivity of free space (ε_0) and hence Coulomb's electrostatic law might vary in time. As mentioned in Sect. 3.3, a similar hypothesis had been proposed by Dicke in 1957. Since the speed of light is given by $c \sim \varepsilon_0^{-1/2}$ it follows that c must be time-dependent. Neither Creer nor Dicke mentioned the $c(t)$ hypothesis, which in other contexts had been suggested since the 1930s and much later would enter cosmological theory. See Kragh (2011), pp. 185–189.

[2] Holmes (1965), p. 35.

[3] Holmes (1965), pp. 983–987, Fairbridge (1964), p. 60.

[4] Carey (1958), p. 316.

[5] MacDougall et al. (1963). See also Dearnley (1965), p. 1286. A similar observation was made independently in Brezhnev et al. (1966).

law discovered by Edwin Hubble in 1929, galaxies at a distance r from us recede at a velocity

$$v = Hr,$$

where H is the Hubble constant. The value $H = 100$ km s^{-1} Mpc^{-1}, widely accepted at the time, translates into $H = 1.03 \times 10^{-4}$ mm year^{-1} km^{-1} and if inserted in the Hubble relation with r equal to the radius of the Earth, the result becomes $v = 0.66$ mm year^{-1}. The Canadian physicists suggested that the "remarkably close agreement between the rate of increase of the Earth's radius and that of the universe" might be worth a closer study. It might be coincidental, they admitted, but it might also indicate some hitherto unrecognized connection between geophysics and cosmology.

The speculation of John MacDougall and his colleagues attracted little attention and was ignored by the astronomers.[6] However, some years later a few scientists took up similar speculations. By the late 1970s Carey had reached the conclusion that "to understand the expansion of the earth, we must seek to understand the expansion of the universe."[7] Not only did he believe that the Earth increased in volume, he now also thought that the mass of the Earth increased according to Dirac's idea of $M \sim t^2$. As mentioned in Sect. 2.2, Dirac had abandoned the hypothesis of spontaneous matter creation in 1938, but when he returned to cosmological theory in the 1970s he also returned to matter creation. Carey wrote: "What then does Hubble's law mean? It means that what I found to apply to the earth also applies to the whole universe—volume expansion and mass increase go hand-in-hand."[8] Carey suggested that terrestrial expansion was due to the formation of new matter in the interior of the Earth, most likely in the form of iron because of its minimum energy per nucleon. He was unable to explain how the iron atoms came into existence and how they gave rise to atoms of other elements— this was another problem he handed over to the physicists.

In a lecture of 1977 Carey speculatively presented a combination of the classical law of gravity and Hubble's expansion law.[9] Without further argument he stated the combined force law in the form

$$F(r) = Gm_1 m_2 \left(\frac{1}{r^2} - \frac{aH^4}{c^4} r^2 \right)$$

Here a is a pure number to be determined by observation and which Carey thought was approximately 10^{20}. As he noted, the fraction $(H/c)^4$ is so small that the

---

[6] But see Klepp (1964).

[7] Carey (1988), p. 328.

[8] Carey (1988), p. 330. Carey's late and highly unorthodox views are summarized in Oldroyd (1996), pp. 275–277.

[9] Carey (1978, 1988, pp. 329–347).

repulsive force corresponding to the last term only becomes significant at very large distances. The factor in front of r^2 is roughly 10^{-44} m^{-4}. There is no need to deal in detail with Carey's law, which lacked any foundation in physics and astronomy, except pointing out that the repulsive term can be seen as corresponding to Einstein's cosmological constant Λ.[10] Indeed, the law has some similarity to the classical modifications of Newton's law that Hugo von Seeliger and a few other astronomers proposed before the emergence of general relativity.[11] Seeliger's modification of 1895 was

$$F(r) = G\frac{m_1 m_2}{r^2} e^{-\Lambda r},$$

where Λ is a small constant and the extra factor corresponds to a repulsive force at very large distances. Seeliger's Λ was a classical analogue of Einstein's cosmological constant.

Carey's excursion into cosmology included what he called the "null universe," which was a version of Edward Tryon's idea of a zero-energy universe given by $GM/R = c^2$ (see Sect. 2.3). But according to Carey the hypothesis of the null universe went farther, for it asserted that "everything in the universe cancels—matter, energy, charge, momentum."[12] Contrary to Tryon, Carey did not associate the idea with the big bang, which he much disliked. He dismissed the explosive origin of the universe as a "myth," a mathematical fabrication with no basis in reality. "Mathematicians revel in harmless sophisticated fantasies, and new-cosmologists buy them as real estate," he quipped.[13] As regards the cosmic microwave background, Carey denied that it was evidence of a big-bang event more than 10 billion years ago. He offered no alternative.

Tryon was the only cosmologist of repute who expressed critical interest in Carey's highly unorthodox views concerning the expanding Earth and the expanding universe. Participating in a symposium on the expanding Earth at the University of Sydney in 1981, Tryon used the opportunity to distance himself from Carey's null universe and other attempts to account for the Earth's expansion by means of matter creation. "It seems very difficult to understand," he said with an understatement, "how matter creation could be the cause of Earth expansion."[14] Tryon further emphasized that the expansion of the Earth, if real, was entirely different from the expansion of the universe. The Sydney symposium was attended by a large number of scientists, although most of them from Australia and with only very few of the attendees being earth scientists of international reputation. Almost all the contributions expressed empirical support for an expanding Earth, whereas

---

[10] Carey (1988) was aware of the formal analogy between the term involving H/c and Einstein's Λ.

[11] For early attempts to modify Newton's law, all of them without involving a time dependence of G, see North (1965), pp. 30–49.

[12] Carey (1976, p. 459, 1983, p. 369).

[13] Carey (1988), p. 332.

[14] Tryon (1983), p. 355.

the physical cause of it was given little attention. The $G(t)$ hypothesis was only mentioned casually.

The triumphant theory of the big-bang universe did not mean an end to the steady-state theory, just as little as the triumphant theory of plate tectonics meant an end to the idea of an expanding Earth. In both cases the rival theories were further marginalized, but they did not quite disappear from the scene of science. In a series of papers starting in 1962 Hoyle and his young collaborator Jayant Narlikar developed new cosmological equations that were modelled on those of general relativity but nonetheless described a modified steady-state universe.[15] In a paper of 1965 Creer referred to the "recent modification of the Einstein theory of relativity" proposed by Hoyle and Narlikar. According to the new theory, he wrote, G would decrease in time, and if "one imagines that half the distant universe could be removed, as if by magic, G would be greater and the earth's orbit would tighten round the sun."[16] However, at the time the Hoyle–Narlikar theory did not include a varying-G hypothesis. It did refer to the kind of magic mentioned by Creer, causing the Earth to be "fried like a crisp" (as Hoyle and Narlikar expressed it), but this was not an effect of G decreasing in time as Creer apparently thought.[17] The two cosmologists only suggested a $G(t)$ theory some years later.

The mathematical framework of the Hoyle–Narlikar theories of "conformal gravity" had a great deal of similarity to that of the Brans–Dicke theory, including that the theories allowed a variety of ways in which G might vary in time. Among the models was one which incorporated several features of the Dirac–Jordan cosmology, including the LNH, creation of matter, and a gravitational constant varying as $1/t$.[18] The temporal variation of G was either

$$\left(\frac{1}{G}\frac{dG}{dt}\right)_0 = -2H_0 = \frac{1}{T},$$

close to the one originally suggested by Dirac, or it might be even stronger, following $G \sim 1/T^2$. The dependence of G on the redshift z followed

$$G(z) = G_0(1+z)^2$$

Hoyle and Narlikar not only discussed the astrophysical consequences of their theory—including the increased luminosity of the Sun in the past, as first mentioned by Teller—but also the consequences for the structure of the Earth. "In principle geophysics could be of decisive importance to cosmology," they wrote in

---

[15] For the series of Hoyle–Narlikar steady-state models, see Kragh (1996), pp. 358–373.

[16] Creer (1965b), p. 39.

[17] Hoyle and Narlikar (1964), p. 204.

[18] Hoyle and Narlikar (1971, 1972). See also Wesson (1978), pp. 38–44, 184, and the review of G (t) cosmologies in Narlikar and Kembhavi (1988).

agreement with the research strategy previously adopted by Jordan and Dicke.[19] The two Cambridge cosmologists suggested that the expansion of the Earth, a result of the decreasing force of gravity, provided the horizontal force that makes the continents drift apart. Based on a three-zone model of the Earth (fluid core, solid core and a thin crust), they obtained for the rate of increase of the Earth's radius

$$\frac{dR}{dt} \cong 10q \, \text{km per } 10^8 \, \text{year},$$

where q is a number about unity.

Based on "discussions we have had with geophysicists" Hoyle and Narlikar claimed that their theory received "powerful support" by solving for the first time the problem of a physical mechanism that could move the continents. This was a reference to Jeffreys' old complaint that physics was unable to explain continental drift, a phenomenon which Jeffreys consequently held to be "impossible."[20] However, by 1972 plate tectonics had largely solved the problem, at least to the satisfaction of most, if not all geophysicists. Jeffreys, a long-time foe of continental drift and defender of the contraction theory, continued to criticize the drift hypothesis for its lack of a physical cause. As late as 1977 Dan P. McKenzie, one of the pioneers of plate tectonics, admitted that a proper dynamical theory for the drifting continents was still missing. The theory was still "kinematic."[21] Yet, although the cause of plate movements was still somewhat unclear, this was no longer seen as a weighty objection to drift and definitely not as an argument for Earth expansion.

On 11 February 1972 Hoyle, serving as President of the Royal Astronomical Society, gave an address on "The History of the Earth" that dealt as much with geophysics as with planetary astronomy. Claiming that there was "a considerable body of evidence" supporting a decreasing G, Hoyle considered the effects on the motion of the Moon, solar luminosity, the temperature of the Earth in the past, and the motion of the continents.[22] All of these subjects had been covered in detail previously, but either Hoyle did not know of the earlier work or he did not care to mention it. He did however refer to the similarity between his own chain of reasoning and that "followed some years ago by Dicke." Indeed, the presidential address had an aura of *déja vu*, or so it must have seemed to Dicke, Jordan and other scientists who for years had grappled with the astro- and geophysical consequences of the $G(t)$ hypothesis.

---

[19] Hoyle and Narlikar (1972), p. 332.

[20] Jeffreys (1924, p. 260) described Wegener's continental drift theory as "an impossible hypothesis ... [unless] forces enormously greater than any yet suggested are shown to be available."

[21] McKenzie (1977), p. 122.

[22] Hoyle (1972). This was not Hoyle's first encounter with paleoclimatology. As early as 1939, in a paper with R. Lyttleton, he suggested that changes in the Earth's climate could be explained as effects of the Sun passing through interstellar clouds of diffuse matter. See Hoyle and Lyttleton (1939).

"With a candid admission that speaking of geophysics was probably 'dangerous in the present company,' he plunged fearlessly into the continental drift debate," as Jon Darius, a London astronomer, reported about Hoyle's lecture.[23] Moreover, "Various eyebrows mounted as Hoyle elaborated on the planetary, stellar, and geophysical effects of a time-varying G." Expressing the revolutionary and holistic spirit at the time in some circles of young scientists, Darius praised Hoyle's lecture for its style rather than its substance. "Hoyle's multidisciplinary approach linking macrocosmic (universe) and microcosmic (planet Earth), may well prefigure the Newthink of future scientists . . . [and point toward] a radical rethinking of scientific attitude and method."

According to Hoyle's calculations, at the time of the formation of the Earth 4.5 billion years ago the solar constant was about five times greater than presently. With β denoting the factor by which the solar constant has increased ($\beta = 1$ at the present), he estimated the absolute temperature of the Earth's surface to have varied as

$$T \cong 280 \beta^{1/4}$$

For $\beta = 3$ or 3 billion years ago, the average temperature on Earth would thus have been 369 K or 96 °C. This he found acceptable given the existence of bacteria living in hot springs at nearly this temperature. Hoyle further sketched a picture of how convection in the mantle combined with expansion could explain how the continental plates were set in horizontal motion. For the rate of increase of the Earth's radius he calculated a value between 6 and 10 km per 10^8 years (that is, of the order 0.1 mm year^{-1}). Hoyle concluded: "The hypothesis of a changing gravitational constant leads one to consider many unusual effects. It does not lead to conflict with available data. It often fits the data better than the conventional theory."[24] The similarity to the view expressed by Jordan some years earlier is striking.

Hoyle's excursion into geophysics was brief and not very convincing. Most earth scientists ignored it and the small camp of expansionists merely noticed the support of the famous cosmologist.[25] Of the 25 citations Hoyle's paper received until 1982, only three were from earth scientists. Hugh Owen, a palaeontologist and cartographer at the British Museum, was among the few who found Hoyle's theory promising. A late advocate of an expanding Earth, he suggested that the Earth's diameter was approximately half its current value 700 million years ago but without offering an explanation of the expansion (Fig. 4.1).[26]

---

[23] Darius (1972).

[24] Hoyle (1972), p. 344.

[25] Hoyle apparently did not consider his geophysical $G(t)$ theory to be important, as he did not mention it in his autobiography. See Hoyle (1994). Carey (1975) mentioned it along with Jordan's theory, but only briefly. As pointed out by Wesson (1973), p. 39, some of the geophysical statements made by Hoyle (1972) and Hoyle and Narlikar (1972) were simply erroneous.

[26] Owen (1976, 1984). Contrary to Carey, who believed that expansion was as old as the Earth itself and had accelerated since the beginning of the Mesozoic, Owen argued for a post-Mesozoic linear expansion.

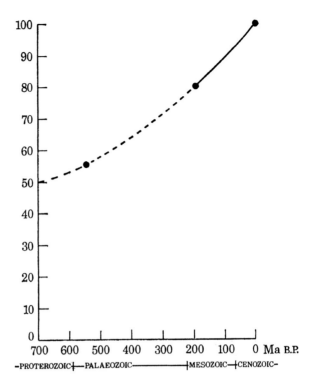

Fig. 4.1 Owen's expanding Earth assuming an exponential growth in the radius through time. *Source*: Owen (1976), p. 230. Reproduced with the permission of The Royal Society

Hoyle was an amateur in geology who came to the subject in his own way, relying on his expertise in astronomy and mathematical physics but with no intention of acquainting himself seriously with the vast literature written by professional geophysicists and geologists. Characteristically, he admitted to be unaware of the relevant geophysical and astronomical data "until I began to hunt in the literature."[27]

Jordan and Dicke were amateurs as well, but in a much more serious manner than the English cosmologist and controversialist. Hoyle's attempt of 1972 to revise the picture of the Earth would not be his last move into territories where he was not at home. Together with his collaborator Chandra Wickramasinghe, Hoyle argued since the mid-1970s that life as well as epidemic diseases came from outer space, and he also suggested a new theory of the ice ages that went flatly against the established view of geologists and climatologists. According to Hoyle, the source of the ice ages should be found primarily in a combination of atmospheric dust and meteorite impacts. Even more disastrous for Hoyle's reputation was his and Wickramasinghe's claim of 1986 that the famous Archaeopteryx fossil (a primitive bird or feathered dinosaur) was a forgery.[28]

---

[27] Hoyle (1972), p. 344.

[28] Hoyle's unconventional ideas are covered in Gregory (2005).

Hoyle's old friend and collaborator, the Cambridge astronomer Raymond Lyttleton, was generally in favour of the steady-state theory of the universe but not of the expanding Earth hypothesis proposed by Hoyle and Narlikar. Contrary to the majority of geophysicists Lyttleton strongly advocated a version of Ramsey's hypothesis of the inner Earth. According to Lyttleton's version, the core was not composed of alloys of iron and nickel, but rather of hydrogen turned to a liquid metal under the very high pressure. He concluded that the primitive Earth was cool and somewhat larger than the present one, perhaps with a radius of 6740 km. Although it might initially have increased in size, for most of its history it had contracted: "A total contraction of 350 km, if spread over say 3.5×10^9 years, would mean only 10^{-2} cm per year, but it is scarcely to be supposed this would take place uniformly and continuously."[29] The contraction of the Earth was held to be responsible for the formation of mountains, something he thought other theories of the Earth were quite incapable of.

Not only did Lyttleton dismiss an expanding Earth, he also dismissed the new and, to his mind, falsely promoted dogma of plate tectonics. Many geophysicists, he wrote, are "inclined strongly to the view that continental drift is occurring, just as there are probably many people still that are inclined to view that 'flying saucers' visit this planet."[30] Lyttleton's ideas about the Earth stood in sharp contrast to those of the mainstream geophysicist Runcorn, who severely criticized the astronomer-turned-a-geophysicist. The disagreement between the two distinguished British scientists evolved into a public controversy.[31]

"For the past 40 years a ghost has been haunting the solar system, if not indeed the universe generally," Lyttleton wrote in a book of 1982. "Its apparition seems first to have been perceived dimly and announced by Dirac (1938), and has led to the notion that the constant of gravitation G may be changing with time."[32] A ghost it was and nothing but, he thought. Based on arguments from astronomy and geophysics, Lyttleton concluded that there was no reason to believe in a varying gravitational constant of either the Dirac–Jordan kind or the Brans–Dicke kind.[33] He also used his model of the two zones of the Earth to criticize the picture of the Earth proposed by Hoyle and Narlikar.

---

[29] Lyttleton (1970), p. 107. Although Lyttleton defended a contracting Earth theory, it was entirely different from the classical thermal-contraction theory based on the assumption of an initially molten Earth.

[30] Lyttleton (1970), p. 120.

[31] See, e.g., *Observatory* **91** (1971): 164–165, *Nature* **240** (1972): 459–460, and *Nature* **241** (1973): 521–523. For Lyttleton's contracting Earth, see also the monograph Lyttleton (1982) and Runcorn's very critical review of it in *New Scientist* (8 September 1983, p. 706). The Lyttleton–Runcorn controversy had no direct bearing on either the expanding Earth or the varying-G hypothesis, for which reason we only mention it.

[32] Lyttleton (1982), p. 177. Lyttleton's critique of plate tectonics and his attempt to revitalize the contracting Earth theory was covered in a supportive article of January 1983 in the British daily *Guardian*. See the quotations in Wood (1985), p. 209.

[33] Lyttleton (1976, 1982, pp. 177–194).

Lyttleton and his collaborator John Fitch saw no reason to support a version of continental drift by means of the varying-G assumption, simply because "by far the greater amount of reliable evidence is entirely in conflict with the hypothesis of drift."[34] In a critical reply to the Hoyle–Narlikar hypothesis they recalculated the two-zone model of the Earth on the basis of Dirac's $G \sim 1/t$, finding a rate of change in radius of $dR/dt \cong 2.3$ km per 10^8 years, or less than a quarter of the value reported by Hoyle and Narlikar. "It is clearly impossible," they concluded, "that a decreasing G could alone cause expansion on such a scale that a fissure between Africa and South America would yawn some 5000 km in width." Based upon astrophysical arguments the Hoyle–Narlikar cosmology was severely criticized also by Jeno Barnothy and Beatrice Tinsley, two American astronomers. They concluded that although Hoyle's and Narlikar's theory was "very ingenious," it "cannot be a valid representation of the observed universe."[35]

Hoyle, Narlikar and Lyttleton were not the only of the small group of steady-state cosmologists who had an interest in the physics of the Earth. According to Thomas Gold, who together with Bondi and Hoyle had pioneered the theory of the steady-state universe in 1948, polar wandering might be the result of a distortion of the figure of the Earth caused by large-scale tectonic movements. He was acquainted with Runcorn and in the mid-1950s participated in meetings with him on polar wandering. While Runcorn at the time was in favour of polar wandering, but not of drifting continents, Gold was more open to the possibility of continental drift, although he did not embrace the hypothesis.[36] Within a few years polar wandering came to be seen as closely related to drift. For example, Runcorn's conversion to drift was to a large extent the result of his study of pole movement through time.

In a joint paper with Bondi, Gold studied the damping of the Earth's nutation, another problem of geophysics. They showed that the damping was due to elastic properties of the mantle and not to motions in the liquid core relative to the mantle.[37] The latter view was widely held and supported by the authority of Jeffreys, but Bondi and Gold claimed that it could not account for the observed damping. These works in geophysics are worth mentioning, not because they were particularly important but because they illustrate that cosmologists were not necessarily foreign to problems of the earth sciences. Contrary to the later entrance of Hoyle in the world of geophysics, the works of Gold and Bondi were strictly separate from the steady-state cosmological model they defended at the time. Many years after Bondi had left research in cosmology he and Lyttleton expressed their disbelief in plate tectonics, a theory they felt lacked incontrovertible support and was plagued by "the absence of an identifiable driving force."[38]

---

[34] Lyttleton and Fitch (1977). See also Wesson (1973), p. 40 and Wesson (1980), pp. 53–54.

[35] Barnothy and Tinsley (1973), p. 349.

[36] Gold (1955), Frankel (2012b), Hallam (1973), p. 40.

[37] Bondi and Gold (1955), Scheidegger (1958), p. 107.

[38] Lyttleton and Bondi (1992), who did not refer to the alternative of an expanding Earth.

4.2   New Creation Cosmologies

Inspired by an old idea due to the German mathematician Hermann Weyl, in 1973 Dirac proposed a modification of Einstein's theory of gravitation that accommodated the Large Numbers Hypothesis (LNH).[39] In a manner somewhat similar to the Jordan–Brans–Dicke theory, in addition to the metric tensor $g_{\mu\nu}$ the new theory required a scalar field to describe the gravitational field. It thus belonged to the family of scalar–tensor theories. Apart from retaining the LNH, the versions of the cosmological theory that Dirac developed after 1973 also retained the $G \sim 1/t$ hypothesis and the idea of two different metrics that he originally had inherited from Milne.[40] But he now returned to spontaneous creation of matter, which he enigmatically described as "a new physical process, a kind of radioactivity, which is quite different from all the observed radioactivity."[41]

According to Dirac matter creation might occur either "additively" (+) or "multiplicatively" (×). According to the first form, matter would be created uniformly through space, and therefore mostly in intergalactic regions; on the other hand, the matter of celestial objects such as the Sun and the Earth would remain essentially constant. Multiplicative creation meant that new matter would be created where it already existed and in proportion to the amount present. Consequently the mass of a star or a planet would grow in the same way as in Jordan's old theory, namely as $M \sim t^2$. Dirac generally preferred multiplicative creation because it clashed less violently with standard general relativity. The idea of this kind of cosmic creation process was not quite new as it had been previously suggested by the Irish-born physicist William McCrea within the context of steady-state cosmology.[42] Table 4.1 lists some of the predictions of the various versions of Dirac's cosmology.

As far as $G(t)$ was concerned, there was no difference between Dirac's two forms, which both led to the rate of decrease

$$\frac{1}{G}\frac{dG}{dt} = -H_0,$$

rather than the three times stronger decrease of Dirac's original theory. With $H_0 = 50$ km s^{-1} Mpc^{-1}, a value advocated by Sandage and widely accepted in

---

[39] Dirac (1973a, b). After having retired from his chair in Cambridge, Dirac moved to Florida, where he joined Florida State University in Tallahassee in 1971. He remained there until his death in 1984.

[40] Dirac (1973c, d, 1974). For a lucid review of Dirac's LNH cosmologies, see Narlikar and Kembhavi (1988).

[41] Dirac (1974), p. 440, Dirac (1978c), p. 12.

[42] McCrea (1964). See also Kragh (1996), p. 359. According to the original steady-state theory proposed in 1948, matter was created uniformly throughout space. While McCrea believed that all new matter was created in galaxies and stars, he did not consider the terrestrial consequences of his hypothesis.

Table 4.1 Predictions of Dirac cosmologies, atomic units

	1938 theory	Later theory (1970s, +creation)	Later theory (1970s, ×creation)
Gravitational constant G	t^{-1}	t^{-1}	t^{-1}
Scale factor R	$t^{1/3}$	t	t
Hubble parameter H	$\frac{1}{3}t^{-1}$	t^{-1}	t^{-1}
Mass of planet	Constant	Constant	t^2
Orbital radius of planet	t	t^{-1}	t
Orbital velocity of planet	t^{-1}	Constant	Constant
Orbital angular momentum	Constant	t^{-1}	t^3

the early 1970s, it gives a variation of -5.1×10^{-11} per year. In Dirac's original theory the scale factor varied as $R \sim t^{1/3}$, whereas in the new theory the universe expanded linearly at least in its later stages. The relation $R \sim t$ was valid for both (+) and (×) creation. This was another point of similarity to Jordan's cosmology and also to Milne's old model, which corresponds to an empty Friedmann universe in standard relativistic cosmology. The change implied a much longer age of the universe, which now came out as approximately equal to the Hubble time or about 18 billion years.

Although Dirac never entered the discussion of an expanding Earth or expressed any real interest in the geological sciences, at a few occasions he casually referred to the consequences his theory might have for the Earth. Thus, in a lecture given in Rome to the Pontifical Academy of Sciences in April 1972 he mentioned, possibly for the first and last time, the expanding Earth. However, he ascribed the expansion to matter creation and not to the decrease of the gravitational constant, as Jordan, Egyed, Dicke and other scientists had done:

> According to the present theory the earth must have been expanding during geological times, owing to the continual creation of new matter inside it. The observed drifting apart of the continents supports this view; also the occurrence of rifts, in the continents and oceans. P. Jordan has written a great deal on this subject.[43]

The expansion of the Earth was not the only consequence of multiplicative creation of terrestrial matter. As Dirac pointed out, spontaneous creation of matter would have a problematic effect on crystal growth over long periods of time:

> It is a little difficult to understand how this [matter creation] can take place in the case of a crystal. Presumably the new atoms must appear on the outside. The rate of multiplication is extremely small, so there is plenty of time for the new atoms to appear in the places most suitable for them. But during the course of geological ages the increase must be quite appreciable, and should be taken into account in any discussion of the formation of crystals in very old rocks. It might lead to insuperable difficulties.[44]

---

[43] Dirac (1973a, b, c, d), p. 11.

[44] Dirac (1974), p. 445. See also Dirac (1973c, d).

There were other difficulties, for how is it that the continuous creation process mysteriously reproduces the chemical composition of matter already existing? If matter were created in some elementary form, for example as protons and neutrons, ancient terrestrial matter as found in old rocks and meteorites would have a very large abundance of hydrogen, contrary to fact.[45] Dirac's remark was taken up by a few scientists, among whom Kenneth Towe, a paleobiologist at the Smithsonian Institution, pointed out that the lattice dimensions of quartz crystals had remained the same over 3 billion years. He consequently concluded that Dirac's hypothesis of multiplicative creation was unacceptable.[46] Runcorn agreed, if for different reasons. According to the British geophysicist, the result of multiplicative creation of matter would be an unrealistic increase in the Moon's mass since its formation by 60 % and a corresponding increase in its radius by 20 %.[47] The large majority of earth scientists agreed with Runcorn's dismissal of Dirac's new theory, or they simply chose to ignore it.

The old problem of the Earth's temperature in the geological past first considered by Teller in 1948 reappeared in Dirac's new theory. If the luminosity of the Sun to varies as $L \sim G^7 M^5$, the combination of $G(t)$ and $M(t)$ implies that

$$L \sim t^3.$$

Combining the increased luminosity with the change in the Earth's orbit caused by the decreasing G parameter, the British astronomer Ian Roxburgh showed that the temperature of the Earth would vary as

$$T_E \sim t^{5/4}$$

Thus the ancient Earth would have been substantially cooler than the present one. "If our understanding of climatology is at all correct," Roxburgh wrote, "this would imply that the Earth was covered with ice and such an ice cover would not have melted even when the equilibrium temperature reached its present value, because of the high albedo of an ice-covered Earth."[48] Apart from Dirac's new theory, varying-G theories in the 1970s included the Jordan–Brans–Dicke scalar–tensor theory, the Hoyle–Narlikar theory, and a new "scale-covariant" extension of general relativity proposed by Vittorio Canuto and co-workers at the NASA Institute for Space Studies.[49] The scale-covariant theory

[45] Eichendorf and Reinhardt (1977).

[46] Towe (1975). See also Gittus (1975) and Van Flandern (1976), p. 53. Vittorio Canuto objected that Towe's reasoning was incorrect since it was based on a wrong application of units. Although Towe admitted his error, he maintained that crystal structure posed a problem for Dirac's theory. See Canuto et al. (1976).

[47] Runcorn (1980).

[48] Roxburgh (1976).

[49] Canuto et al. (1977), Canuto and Lodenquai (1977).

was consistent with Dirac's theory and accommodated the LNH. For our purpose it can be considered an alternative and mathematically more sophisticated way of interpreting Dirac's cosmological ideas, including a time-variation of G given by $dG/Gdt \cong -6 \times 10^{-11}$ per year. Although Canuto's theory allowed G to vary, it did not require it.

As Hönl and Dehnen in 1968 had criticized Jordan's cosmology because it did not agree with the spectral shape of the observed microwave background radiation, so Dirac's new cosmology faced the same problem. Because of the continuous creation of photons proportional to t^3 in Dirac's new theory, it was unable to account in a natural manner for the blackbody spectrum of the cosmic microwave background. Dirac noted the problem in his Rome address of 1972:

> The increase in the number of photons must apply also to the microwave radiation that is observed falling continuously on the earth and is believed to come from a primordial fire-ball. ... With the number of photons increasing, as required by the present theory, the black-body character is not preserved. ... It would be against the present theory if the radiation was observed to be black at the present time, as that would involve an unjustifiable coincidence.[50]

At the time Dirac did not consider it a serious problem, simply because he was not convinced that the wavelengths of the microwave background were really distributed as required by Planck's radiation law. He thought that the radiation might be "grey" rather than black.[51] However, new measurements of the background radiation over a larger range of wavelengths made it increasingly difficult to deny that the radiation was in fact of the blackbody type. By 1975 measurements of the microwave background included data in the range from $\lambda = 0.05$ cm to $\lambda = 74$ cm, and they all fitted the blackbody spectrum very precisely.

Dirac's observation that his cosmology with creation of photons was inconsistent with the Planckian form of the microwave background was taken up and amplified by the Yale astrophysicist Gary Steigman in papers highly critical to the LNH. According to Steigman, in the early phase of Dirac's LNH universe there was not sufficient time for any nuclear reactions to proceed and hence no explanation of the large abundance of helium.[52] Moreover, there could be no thermodynamic equilibrium between photons and electrons, meaning that also the blackbody-shape of the background radiation was without explanation. Contrary to this situation, the standard big-bang theory explained naturally and in detail the microwave background and the helium abundance. Noting that "The application of the LNH to physics and cosmology is fraught with ambiguity," Steigman expressed

---

[50] Dirac (1973d), p. 13.

[51] Contrary to a black body a "grey body" does not absorb all incident radiation and it emits less total energy than a black body. For this reason the radiation from a grey body does not satisfy the fundamental Planck distribution law.

[52] Steigman (1976, 1978).

surprise of "the recent explosion of interest" in cosmological models of this kind.[53]
To the mind of Steigman, Dirac cosmologies were scarcely worth taking seriously.

It took a couple of years until Dirac became convinced that the cosmic background was in fact blackbody-shaped and then decided to revise his cosmological theory. At a conference held in the University of Miami in early 1975 he came up with an alternative explanation of the cosmic background quite different to the standard interpretation in terms of an original decoupling of matter and radiation some 380,000 years after the big bang event at $t = 0$. With $M =$ the proton's mass and $k =$ Boltzmann's constant Dirac considered the large dimensionless ratio

$$\frac{Mc^2}{kT} = 4 \times 10^{12} \quad \text{or} \quad \left(\frac{Mc^2}{kT}\right)^3 \cong 10^{38}.$$

From this ratio and the LNH Dirac inferred that the temperature T of the microwave background should vary as

$$T \sim t^{-1/3} \quad \text{rather than} \quad T \sim t^{-1}.$$

With the electron's mass m instead of the proton's mass M, the variation would become $T \sim t^{-1/4}$. In any case it would be much slower than determined in the conventional theory. To resolve the discrepancy Dirac hypothesized that the present radiation originated not in a primeval "fireball," but from a recent decoupling of photons from a hypothetical intergalactic medium. He described the medium as "intergalactic ionized hydrogen, sufficiently tenuous not to interfere with ordinary astronomical observations, yet sufficiently dense ... to take on essentially the temperature of the intergalactic gas."[54] With this assumption the temperature dependence of the radiation would have been $t^{-1/3}$ (or $t^{-1/4}$) until recently, when it changed to the ordinary t^{-1} as a result of the decoupling.

The hypothetical intergalactic medium remained hypothetical. Nonetheless, Dirac stuck to his theory and felt justified to conclude not only that it could account for the observed radiation background but even that "The microwave radiation thus provides *confirmation* of our present picture."[55] Whether based on additive or multiplicative creation, in the high-frequency region the radiation curves derived by Dirac differed somewhat from the blackbody spectrum. Thus there was in principle a way to decide experimentally if Dirac's alternative was correct or not. "I have heard that plans are being considered for making microwave observations

---

[53] Steigman (1978). See also Wesson (1980), pp. 15–17.

[54] Dirac (1975), p. 452.

[55] Dirac (1979), p. 23. Emphasis added. Dirac's rather ad hoc mechanism for the cosmic microwave background was not the only attempt in the period to explain away its origin in a fireball caused by the big bang. In 1967 Hoyle and Wickramasinghe claimed that the radiation could be understood on the basis of the steady-state theory, namely as starlight thermalized by interstellar grains. See Kragh (1996), p. 356.

from a space platform," Dirac said in his 1975 address. "Thus we may hope some day to have an experimental check on this theory and an independent way of deciding between the two kinds of creation."[56]

At a symposium in Tallahassee, Florida, in the spring of 1978 Dirac returned to matter conservation, much as he had done 40 years earlier. Referring to particle creation following $N \sim t^2$, he now said: "I have been working with this assumption of continuous creation of matter for a number of years, but find difficulties in reconciling it with various observations, and now believe it should be given up."[57] In his new theory there was no genuine creation of matter, but only an increase in the amount of matter in the observable region of the universe. Space was flat and density and expansion followed the relations

$$\rho \sim t^{-1} \quad \text{and} \quad R \sim t^{1/3}.$$

As in the earlier versions of the theory, G varied as t^{-1}. As to the cosmic microwave background, Dirac maintained his hypothesis of an interstellar medium.

The problem of the background radiation was also considered by Canuto and S.-H. Hsieh, according to whom it was far from fatal to Dirac's theory including matter creation.[58] Not only did they argue that the spectral composition of the radiation was consistent with the LNH and Dirac's theory, they also claimed that it followed naturally from the scale-covariant version of it. On the other hand, it was argued against the LNH-based theory of Canuto and Hsieh that it did not provide enough time in the early hot universe to produce large amounts of helium, for which reason "this cosmological model contradicts observation." Canuto and Hsieh denied that this was the case.[59] Whereas Canuto found Dirac's reasoning with regard to the microwave background to be acceptable, other physicists did not. Victor Mansfield at Colgate University, New York, showed that there were grave difficulties with Dirac's alternative explanation of the background radiation.[60] Not only was the explanation ad hoc and the intergalactic medium hypothetical, the medium would also have to have properties (electron density and temperature) that squarely contradicted observations in extragalactic astronomy.

The gravitation theory proposed by Canuto led to astrophysical and geophysical consequences which in some cases were the same as in Dirac's theory, but in other cases were different. Based on the LNH Canuto and his co-workers concluded that,

---

[56] Dirac (1975), p. 454. The COBE (Cosmic Background Explorer) project was initiated in 1972 but the satellite was launched into orbit only 17 years later. The data from COBE proved that the background radiation fitted a blackbody curve of temperature 2.73 K most perfectly.

[57] Dirac (1978a), p. 170, Dirac (1979). The 1978 symposium in Dirac's honour marked the fiftieth anniversary of the Dirac wave equation in quantum mechanics. Among the participants were several leading quantum physicists, including Eugene Wigner, Freeman Dyson, Murray Gell-Mann, Gerard 't Hooft, and Frank Wilczek.

[58] Canuto and Hsieh (1978).

[59] Falik (1979), Canuto and Hsieh (1980b).

[60] Mansfield (1976).

if matter creation were assumed, the radius of the Earth would increase at a rate from 0.2 to 0.3 mm per year.[61] Without matter creation the result would be smaller by a factor of ten. However, the scale-covariant theory also allowed different variations, including that a larger G in the past implied a larger Earth radius. In a later paper Canuto suggested that, for a constant mass of the Earth, the radius of the young Earth might have been slightly *greater* than today.[62]

Canuto and his collaborators also obtained new results when re-analysing the problem which Teller had first addressed in 1948, that is, the effect of $G(t)$ on the luminosity of the Sun and the past temperature of the Earth (see Sects. 2.5 and 3.5). According to Teller the luminosity of the Sun varied as $L \sim G^7 M^5$, which was the cause of the high heating of the Earth with a larger G in the past. Canuto and Hsieh argued from standard general relativity that the product GM must be a constant, and from this they derived that "the Sun's luminosity is constant in time, independently of whether G varies or not."[63] When the change in the Sun-Earth distance caused by $G(t)$ was taken into account, it was found that the past temperature of the Earth agreed with paleontological data.

Paul Wesson, a specialist in alternative theories of gravity, was impressed by Canuto's scale-covariant theory. It yields, he said, "as good (or better) agreement with all of the standard cosmological tests as does Einstein's general relativity."[64] However, geophysicists paid even less attention to Canuto's theory than they did to the Hoyle–Narlikar theory. The 1977 paper of Canuto and his group was cited 49 times in the period 1977–1982, and none of the citations appeared in journals principally devoted to the earth sciences.

A result somewhat similar to Canuto's had been found a few years earlier by Chao-Wen Chin and Richard Stothers in an examination of Dirac's theory with multiplicative creation of matter.[65] In this case, although the greater G in the past would increase the Sun's luminosity, its smaller mass would have the opposite effect and the net result would be an energy output of approximately the same amount as in the standard model of the Sun. The two American space scientists concluded that the multiplicative (but not the additive) version of Dirac's theory was consistent with known facts about solar physics. They added a potential problem for the creation hypothesis, a fossil analogue of the crystal problem first mentioned by Dirac: "But why are well-preserved Precambrian and early Cambrian fossils in essentially perfect shape if their masses have increased by a significant percentage?"

Yet another examination of Dirac cosmologies was undertaken by André Maeder, an astrophysicist at the Geneva Observatory in Switzerland who in 1977

---

[61] Canuto et al. (1977).

[62] Canuto (1981).

[63] Canuto and Hsieh (1980a). See also Canuto and Lodenquai (1977), which contains a derivation of the Earth's temperature in the past assuming both $G \sim 1/t$ and $M \sim t^2$.

[64] Wesson (1980), p. 35.

[65] Chin and Stothers (1975).

arrived at about the same result as Chin and Stothers with regard to the effect of solar radiation on the surface temperature of the Earth: the Dirac model with ($\times$) creation did not lead to the same difficulties as the original no-creation $G \sim 1/t$ model. Moreover, Maeder concluded that the multiplicative model actually led to a significantly better agreement with the measured neutrino flux from the Sun than the standard solar model assuming $G = \text{constant}$. It thus appeared to promise a solution to the much-discussed solar neutrino problem. This problem, first noticed in the late 1960s, was that the measured solar neutrino flux was significantly smaller than the one predicted by the standard model of the Sun's interior. While Raymond Davis and associates in 1968 found an upper bound on the neutrino flux of 3 SNU, calculations based on the standard solar model resulted in approximately 7.5 SNU.[66]

The conflict between measurements and theoretical predictions persisted for three decades until it was resolved with the discovery of neutrino oscillations at the end of the 1990s, implying that neutrinos carry mass and can change between the electron form ν_e and the muon form ν_μ. While Maeder rejected Dirac's (1938) model and the later ($+$) creation model, he optimistically concluded that "the non-conservative case with past variable G and mass gives results in extremely good agreement with reality!"[67]

In papers from the early 1980s the Japanese physicist Shin Yabushita from Koyoto University investigated the relationship between Dirac's $G(t)$ hypotheses and an expanding Earth. On the assumption of an equation of state for a three-zone model with liquid core and Dirac's $G \sim 1/t$ hypothesis with mass conservation he found a result agreeing with the one obtained by Lyttleton and Fitch, namely

$$\frac{1}{R}\frac{dR}{dt} = -0.062\frac{1}{G}\frac{dG}{dt}.$$

From Dirac's new hypothesis with multiplicative creation of matter Yabushita deduced

$$\frac{1}{R}\frac{dR}{dt} = -0.61\frac{1}{G}\frac{dG}{dt}.$$

With the Hubble time $H = 6 \times 10^{-11}$ year^{-1} this gives

$$\frac{dR}{dt} = 7 \times 10^{-3} \text{ cm year}^{-1} \quad \text{(no creation)}$$

and

---

[66] One SNU (solar neutrino unit) equals 10^{-36} nuclear interactions per target atom per second. At the age of 88, in 2002 Davis was awarded the Nobel Prize in physics for his pioneering work on solar neutrinos.

[67] Maeder (1977), p. 366.

$$\frac{dR}{dt} = 2.5 \times 10^{-2}\,\mathrm{cm\,year^{-1}} \quad (\times \text{ creation}).$$

Since Yabushita took the average rate of terrestrial expansion to be 0.48 mm per year, he concluded that "if the rate of radial expansion in the near past or the radius at the time of Earth formation is well established, it will give a strong support for cosmologies with variable G and creation."[68] Of course, the problem was that the expansion was far from well established. In a subsequent paper Yabushita adopted Ramsey's phase-change hypothesis of the inner Earth and now came to the opposite conclusion: "The multiplicative creation appears to contradict the Earth expansion as claimed by the geologists."[69]

4.3 Testing Varying Gravity

The $G(t)$ hypothesis was considered relevant for early life on Earth and occasionally it also entered into other fields of paleobiology. The first time might have been in 1971, when Bronisław Kuchowicz, a Polish physicist and radiochemist, pointed out that according to Jordan's varying-G theory the gravitational acceleration g on the surface of the Earth might have been twice as large in the Palaeozoic as it is now (that is, $g \sim 20$ m s^{-2}). Kuchowicz suggested that the Dirac–Jordan hypothesis was potentially relevant to biologists and palaeontologists. Among other things, he speculated that "the rather clumsy shapes of the first land animals, in the periods when the gravity was larger than now, [might] be at least in some part related to the relatively larger weight, tending to squash these organisms."[70] He may have thought of animals such as the mammal-like therapsids. Kuchowicz's appeal to the biologists fell on deaf ears, but ten years later an American zoologist wondered why the largest known land mammal was the *Baluchitherium* of body mass approximately 20 tons.[71] He argued that this was an upper limit imposed by gravity to the size of land mammals. In a postscript he referred to Dirac's hypothesis, although pointing out that in this particular case the effect of a larger gravity in the past was negligible. *Baluchitherium* lived in the Oligocene, only 30 million years ago and thus at a time when the surface gravity was practically the same as today.

This kind of reasoning, that the weight of prehistoric land animals indicated an upper limit to paleogravity, was not new. Alexander Stewart, a geologist at the University of Reading, referred to the thirty tons heavy *Apatosaurus*—better known as *Brontosaurus*—as evidence that in the Upper Jurassic some 150 million years ago gravity could not have been much greater than today, perhaps $g \leq 1.2\,g_0$.[72]

---

[68] Yabushita (1982), p. 141.
[69] Yabushita (1984), p. 45.
[70] Kuchowicz (1971), who cited Jordan (1955, 1966).
[71] Economos (1981).
[72] Stewart (1977). See also Stewart (1970).

Several later expansionists have taken up similar arguments to explain the great size and weight of some of the dinosaurs, but with different conclusions. According to Stephen Hurrell, an author and engineer, the size of the large dinosaurs can only be understood on the assumption that the Earth was expanding and gravity had been much *weaker* in the past.[73] A few other authors, but no professional scientists, subscribe to the increasing-gravity theory and its implications for the evolution of large animals.

In the two decades after 1970 much work was done on testing a possible decrease of the gravitational constant. The LNH conjecture and related ideas motivated a host of experimental and observational studies.[74] A few laboratory tests were performed in the period, but their accuracy was limited to the level $10^{-7} < dG/G < 10^{-8}$, which made them irrelevant with respect to the theoretical predictions of the order 10^{-10} or 10^{-11} per year.

While geophysical and paleontological tests provided some corroborating evidence for the $G(t)$ hypothesis it was generally realized that tests of this kind were unable to deliver decisive proof for or against the hypothesis. There were simply too many uncertain factors involved, such as Dicke had pointed out in the mid-1960s. For example, examinations of daily, monthly and annual growth rings of fossil bivalves and corals suggested that the number of days in the year was greater in Devonian times than at present. Other studies using the "paleontological clocks" or "speaking stones" indicated that the number of days per lunar month and per year had decreased significantly since the Ordovician era.[75] Similar reasoning had been applied by Wells in his early attempt to establish a geochronology independent of geophysical and astronomical data, but at the time with little success. The more recent data resulted in paleontological evidence for the slowing down of the Earth's rotation in the past, which might possibly, but only possibly, be an effect of decreasing gravity.

The clever method of paleontological clocks led to the result that

$$\frac{1}{G}\frac{dG}{dt} = (-0.5 \pm 2) \times 10^{11} \ \text{year}^{-1}$$

According to G. M. Blake, the result ruled out Dirac's early version of the LNH and also his cosmology with additive creation of matter.[76] On the other hand, the result and the method on which it was based showed that with the hypothesis of multiplicative creation Dirac's $G(t)$ was in "almost as close agreement" with fossil data as the conventional $G = $ constant hypothesis.[77] Still, the growth ring method was unable to come up with an answer to whether gravity varies or stays constant.

---

[73] Hurrell (2011). See also Scalera et al. (2012), pp. 307–366.

[74] Gillies (1983, 1997) are detailed reports on gravitational studies.

[75] Pannella (1972).

[76] Blake (1977, 1978).

[77] Blake (1978), p. 405.

Fig. 4.2 Ancient values of the gravitational constant G and the surface gravity acceleration g on the assumption of Dirac's hypothesis and the expanding Earth hypothesis. *Source*: Stewart (1977). Reproduced with the permission of The Geological Society of London

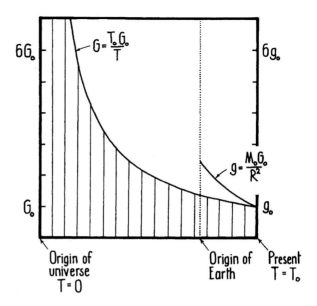

    The verdict from other lines of geological and paleontological inquiry was hardly more promising. For example, Stewart focused on the gravitational acceleration g at the surface of the Earth rather than on G.[78] The two quantities are related as

$$g = G\frac{M}{R^2},$$

where M denotes the mass of the Earth and R its radius. Thus, if the Earth expands as the result of a weakening G, the variation in g will be stronger than the variation in G (Fig. 4.2). From a variety of paleontological evidence, ranging from the permanent existence of an atmosphere since the Archaean to the mineralogical composition of ancient clays, Stewart concluded that

$$g_0 \leq g(\text{Precambrian}) \leq 1.9 g_0$$

and

$$g_0 \leq g(\text{Phanerozoic}) \leq 1.4 g_0.$$

Stewart concluded that the data "flatly contradict rapid Earth expansion models" such as those preferred by Carey and Owen.[79] This kind of accelerated expansion, which was also favoured by van Hilten, can be represented by the expression

---

[78] Stewart (1977, 1978, 1983).

[79] Stewart (1978), p. 155.

$$R(t) \cong R_0[0.6 + 0.4\exp(-kt)].$$

R_0 is the present radius of the Earth, t is the time in the past ($t = 0$ at present) and k is an empirical constant of the order 5×10^{-9} year^{-1}. The average expansion is about 1 cm year^{-1} and the primeval Earth is assumed to have had a radius of 0.6 $R_0 = 1700$ km. In the Archaean, some 3.7 billion years ago, the data analysed by Stewart suggested the limit $g \leq 1.2\, g_0$, which corresponds to $dG/Gdt \leq 2.4 \times 10^{-10}$ year^{-1} under the assumption of $M = $ constant. This was a fairly sharp constraint, but not quite sharp enough to test the cosmological models assuming $G(t)$. Stewart noted that models of slow expansion of the Egyed type survived his test, if only barely so.

After many years of research there still was no unambiguously positive evidence from the earth sciences that the force of gravity was stronger in the past. One could still maintain that "the hypothesis of a decreasing gravitational constant is consistent with geophysical phenomena,"[80] but this was not enough. Consistency between a hypothesis and a range of phenomena within the domain of the hypothesis does not imply that the hypothesis is either confirmed or disconfirmed. It was left to the astronomers to test more accurately the predictions of the various $G(t)$ hypotheses.

The astronomical methods were essentially based on measurements of the distances to the Moon and the nearby planets over long periods of time. Using the technique of radar-echo time delays, in 1971 Irving Shapiro and collaborators had collected sufficient data for the distances to Mercury and Venus to conclude that G varied less than 4 parts in 10^{10} years. In other words, "our result shows no evidence for a time variation of the gravitational constant."[81] A few years later Shapiro and his group improved the constraint to

$$\left| \frac{1}{G}\frac{dG}{dt} \right| < 1.5 \times 10^{-10}\,\text{year}^{-1}.$$

Other methods based on astrophysical reasoning sharpened the results obtained by the radar-echo method. If G decreases a cluster of particles will expand and some of the particles escape from the cluster. Astrophysicists David Dearborn and David Schramm at the University of Texas, Austin, pointed out that the "particles" might be either galaxies or stars (as in globular clusters) and that the study of such celestial objects might give information about the possible variation of G. Their calculations resulted in a limit even stronger than the one found by Shapiro, namely,

---

[80] Narlikar and Kembhavi (1988), p. 484.

[81] Shapiro et al. (1971).

$$\frac{1}{G}\frac{dG}{dt} \cong -5 \times 10^{-11}\,\text{year}^{-1}.$$

While the result did not rule out the Brans–Dicke theory, Dearborn and Schramm concluded that it was inconsistent with Dirac's $G(t)$ variation and also with the cosmology proposed by Hoyle and Narlikar.[82]

It seemed that the heavens conspired against the hypothesis of a decreasing gravitational constant no less than the Earth did. But in 1975 Thomas Van Flandern at the U.S. Naval Observatory in Washington, DC created a minor sensation by announcing a positive and apparently reliable result. *New Scientist* joyfully commented that the decrease in gravity was so very slow that "it will not assist slimmers: a person weighing about 10 stones would lose one millionth of the weight of a paper clip each year!"[83] Too bad! In view of Jordan's early prize essays to the Gravity Research Foundation (GRF), it is worth noting that in 1974 Van Flandern submitted an essay to GRF, essentially a summary draft of the paper published the following year. The only major difference was the value of the variation of G, which he gave as $dG/Gdt = (-11 \pm 3) \times 10^{-11}$ year^{-1}. The essay won a second prize.[84]

From a careful analysis of data for the orbit of the Moon Van Flandern found a value for the secular deceleration of its mean longitude that could not be fully assigned to known geophysical causes such as tides. He proposed that the residual was due to a decreasing G and concluded that there is "a secular decrease in the Universal Gravitational Constant, G, at the rate of $(-8 \pm 5) \times 10^{-11}$/year."[85] Although he noted that the conclusion was "not compelling," he found it to be "strongly indicated, since the anomalous accelerations almost certainly have a cosmological cause, and since there is other supporting evidence." As possible alternatives to the $G(t)$ assumption Van Flandern mentioned the hypothesis that space might be expanding uniformly even in solid bodies and small-scale systems. This kind of alternative was developed by Wesson, but he and Van Flandern realized that it was speculative and without experimental support.[86]

Among the supporting evidence Van Flandern referred to the geophysical and astronomical evidence discussed by Dicke and other scientists, including paleoclimatology and an expanding Earth. Van Flandern further pointed out that the data were consistent with the varying-G theories proposed by Dirac, Brans and Dicke, and Hoyle and Narlikar. In a later article he added to the list Canuto's scale-

---

[82] Dearborn and Schramm (1974).

[83] *New Scientist* (15 May 1975), p. 364.

[84] See http://www.gravityresearchfoundation.org/pdf/awarded/1974/vanflandern.pdf. The first prize of 1974 went to the British cosmologist Joseph Silk. On Jordan and the GRF, see Sect. 3.2.

[85] Van Flandern (1975a), p. 339.

[86] Wesson (1980), pp. 71–72.

covariant theory, within the framework of which the value of the relative change in G would be $(-6.4 \pm 2.2) \times 10^{-11}$ per year.[87]

Van Flandern first presented his result in 1974, in a communication to the American Astronomical Society and also in an address to the Seventh Texas Symposium on Relativistic Astrophysics that convened in Dallas at the end of the year.[88] In the first case he reported as the best estimate $dG/Gdt = (-9 \pm 3) \times 10^{-11}$ year^{-1} and in the second case $dG/Gdt = (-7.5 \pm 2.7) \times 10^{-11}$ year^{-1}. In Dallas he emphasized the cosmological significance of the result, noticing how close the numerical value was to the Hubble expansion rate, which Allan Sandage and his Swiss collaborator Gustav Tammann had recently determined to $H = 55 \pm 7$ km s^{-1} Mpc^{-1} or $(5.6 \pm 0.7) \times 10^{-11}$ year^{-1}. Van Flandern spoke out in favour of "the almost inescapable conclusion that the observed universe is more like that predicted by decreasing gravitational constant than by the usual model."[89] At a workshop in Tallahassee in November the following year Van Flandern recalculated his data according to Dirac's theory with ($\times$) creation, presenting the result as

$$\frac{1}{G}\frac{dG}{dt} = (-5.8 \pm 3.1) \times 10^{-11} \text{ year}^{-1}.$$

He found the agreement with Hubble's constant to be impressive and an argument in support of Dirac's cosmology.[90] Another participant at the workshop, the British physicist Paul Muller, agreed. "It is concluded," Muller said, "that there are strong reasons for accepting the cosmological $\dot{G}/G$ [$=dG/Gdt$] found from the observations, and consistent with the Hubble constant." Muller thought that Van Flandern's results were "very encouraging" and "very strong."[91] But it soon turned out that Muller's optimistic evaluation was unfounded.

At the time Van Flandern was completing his 1975 article, Dirac gave a talk in Sydney, Australia, about his new versions of varying-G cosmology. Referring to Van Flandern's value $(-8 \pm 5) \times 10^{-11}$ year^{-1} he pointed out that although it agreed with the "primitive theory" ($G \sim 1/t$ without matter creation) it did not agree with Dirac's recent ideas. On the assumption of additive matter creation the result derived from Van Flandern's data would be $(16 \pm 10) \times 10^{-11}$ year^{-1} whereas multiplicative creation gave $(-16 \pm 10) \times 10^{-11}$ year^{-1}. Even though the latter result compared much better to the theoretical value -6×10^{-11} year^{-1} than the former one, it still disagreed with it. "Van Flandern has been continually checking and rechecking his calculations and has been modifying his results somewhat," Dirac said. "His most recent results that I have heard about are

---

[87] Van Flandern (1981).

[88] Van Flandern (1974, 1975b).

[89] Van Flandern (1975b), p. 495.

[90] Van Flandern (1978).

[91] Muller (1978), p. 113.

considerably less than his original [that is, $(-8 \pm 5) \times 10^{-11}$ year^{-1}] and they are getting closer to what the theory wants."[92] Dirac was optimistic.

As Van Flandern made clear in a subsequent article in *Scientific American*, he thought that observational evidence favoured Dirac's $G(t)$ hypothesis including multiplicative creation of new matter. Apart from his own data, which he now gave as $dG/Gdt = (-7.2 \pm 3.7) \times 10^{-11}$ year^{-1}, he referred to "the large rift faults in the crust of the Moon and Mars," suggesting that they might be the result of the expansion of the two celestial bodies since they were formed. Expansion, he wrote, could also help to explain "how a continent such as Antarctica can be almost surrounded by a mid-oceanic ridge and yet be apparently drifting away from the ridge at the same time."[93] In a review of Wesson's *Cosmology and Geophysics* he suggested that expansion might be a regular feature of planetary evolution and not a phenomenon confined to the Earth only.[94]

Realizing that ex nihilo creation of matter was unpalatable to most physicists, Van Flandern suggested an alternative mechanism based on the no less heterodox idea of gravitational shielding, for instance in the form that the gravitational field is absorbed in the intervening medium.[95] In the years to come he and a few other scientists (as well as non-scientists) would cultivate this and similar unorthodox ideas, thereby adding to their estrangement from mainstream physical science.[96]

Not unlike Hoyle, but even more extremely, by the early 1980s Van Flandern became increasingly dissatisfied with mainstream science and drifted into areas which were decidedly speculative and non-mainstream, not to say pseudo-science. He promoted the belief that the major planets sometimes explode, that gravity propagates with a speed billions of times greater than light, and that there had been intelligent life on the Moon. He also turned into a vocal opponent not only of big-bang cosmology but also of expanding universe models, which he claimed were inferior to models of the static universe. It was Van Flandern's experience with testing Dirac's $G(t)$ hypothesis that first made him critical of the norms and practices of the scientific community, and eventually deciding to separate from it. As he wrote, "This experience led me to realize how fragile were the assumptions underlying the Big Bang and other theories of cosmology, when even the constancy

---

[92] Dirac (1978b), p. 84.

[93] Van Flandern (1976), p. 51.

[94] *Nature* **278** (1979): 821.

[95] The idea of gravitational shielding or absorption has a long history. The Italian physicist Quirino Majorana (an uncle to the better known physicist Ettore Majorana) performed in the early twentieth century delicate experiments to prove that matter is not transparent to the gravitational flux. Although he came up with a positive result, it was generally disbelieved. Today it is agreed that there is no such thing as gravitational absorption.

[96] See for example Van Flandern (2002), an extensive argument for a classical, finite-range "graviton" model of gravitation. In his later years Van Flandern published many of his articles in *Meta Research Bulletin*, a journal he established as a vehicle for his own ideas and other non-conventional science. For a collection of his many theories belonging to this class of science or pseudo-science, see Van Flandern (1993).

of gravitation ... had been called into question." He described the attitude of most of his colleagues, namely to deny that G could possibly vary, as "understandable, but unscientific."[97]

Van Flandern seems to have had second thoughts about his result announced in 1974. At the 1978 Tallahassee conference Dirac reported that he had recently received a letter from Van Flandern in which "he says that in the intervening two years the accuracy of his result has not increased in the expected way and he is wondering whether there is not some undiscovered systematic error."[98] Four years later, at a conference celebrating the centenary of Einstein's birth, 79-year-old Dirac again suggested that Van Flandern's results "are still not conclusive."[99] He advised to wait "a little longer." Indeed, Van Flandern's belief that he had found solid evidence for a decreasing G turned out to be premature.

Although Wesson found Van Flandern's result to be valid,[100] his data analysis was criticized by other astronomers who were not convinced that he had fully taken into account the uncertainties in the influence of tidal forces on the Earth-Moon system. By the early 1980s his result was generally seen as anomalous and hardly reliable. Although it had not been shown to be wrong, it had not been confirmed either and stood in contradiction to a wide range of other determinations of the possible limits of varying gravity. The case for $G(t)$ "remains unproven" as the theoretical physicist Freeman Dyson diplomatically expressed it in a critical comment at the Tallahassee conference.[101]

Nonetheless, in 1981 Van Flandern suggested that his non-zero result was concordant with preliminary data from planetary radar ranging and lunar laser ranging. Whereas a few years earlier he had supported the ($\times$) version of Dirac's cosmology, he now switched to the ($+$) version, on the basis of which model he reported his most recent value to be $dG/Gdt = (-3.2 \pm 1.1) \times 10^{-11}$ per year. Van Flandern related his belief in $G(t)$ to the old distinction between atomic and dynamical time scales going back to Milne's conventionalist philosophy. He spelled out its cosmological consequences in a manner strikingly similar to what Milne had done 40 years earlier:

> This new result therefore implies that the big bang singularity can occur in atomic processes, but not necessarily in dynamical processes! While the universe has an "age" of about 10 years in atomic time, its "age" in dynamical time must be much greater, and perhaps infinite. Obviously this idea requires much additional elaboration which is beyond the scope of this paper. It does, however, represent a curious wedding of the big bang and the steady state theories of cosmology which contradicts neither, but would considerably modify our intuitive understanding about the possible origin and ultimate fate of the universe.[102]

---

[97] Van Flandern (1993), p. xix.

[98] Dirac (1978a), p. 174.

[99] Dirac (1982), p. 88.

[100] Wesson (1978), p. 166.

[101] Dyson (1978), p. 167.

[102] Van Flandern (1981), p. 815. See also Van Flandern (1976), p. 52.

In any case, the optimistic conclusion about a varying G was soon challenged by more precise data from the Viking landers on Mars. Based upon more than one thousand measurements of the Earth-Mars distance determined between 1976 and 1982, Ronald Hellings and collaborators concluded that

$$\frac{1}{G}\frac{dG}{dt} = (0.2 \pm 0.4) \times 10^{-11} \, \text{year}^{-1},$$

a result that contradicted Dirac's variation $G \sim 1/t$.[103] Interestingly, Hellings' team included Canuto, who until that time had been a proponent of the LNH and varying-G theories. Commenting on the paper by the American team, *New Scientist* spelled out its consequence in plain words: "Gravity does not vary in time."[104] Eighty-one-year-old Dirac was informed of the result, but without it shook his belief that G decreases in time in agreement with the LNH. He believed that the experiments must be wrong, not the theory.[105]

The astronomical verdict of zero change received further support, should such be needed, from the planetary sciences and geophysics. According to several researchers the Moon and Mercury were better testing grounds for gravity-induced expansion than the Earth. Canadian scientists D. J. Crossley and R. Stevens at the Memorial University of Newfoundland argued that Mercury's radius would have increased by around 100 km if the Hoyle–Narlikar theory were correct.[106] They found no evidence for expansion and hence also no evidence for a decreasing gravitational force.

Michael McElhinny and his two collaborators S. Taylor and David Stevenson at the Australian National University in Canberra focused on the Moon. From an analysis of lunar data, some of them obtained from the Apollo Programme, they concluded that for nearly 4 billion years the radius of our satellite had changed by less than 1 km or 0.06 %. For the Earth they used paleomagnetic data to infer that the radius of our globe cannot have expanded more than 0.8 % over the past 400 million years. The relationship between $R(t)$ and $G(t)$ can in general be written as

$$\frac{1}{R}\frac{dR}{dt} = -\frac{\alpha}{G}\frac{dG}{dt},$$

where α is a quantity depending on the equation of state of the Earth. By calculating α in terms of pressure and density an estimate for the rate of $G(t)$ can be obtained. McElhinny and collaborators found $\alpha = 0.085 \pm 0.02$ for the Earth and $\alpha = 0.004 \pm 0.001$ for the Moon. Recall that Dicke had suggested $\alpha = 0.1$ for the

---

[103] Hellings et al. (1983).

[104] *New Scientist* (17 November 1983), p. 494.

[105] See Kragh (1990), p. 354.

[106] Crossley and Stevens (1976).

Earth.[107] Considering not only the Earth and the Moon, but also Mars and Mercury, the Australian group concluded that "the upper limit of 8×10^{-12} year^{-1} for the rate at which G decreases in a constant mass theory is realistic."[108] The value ruled out the Hoyle–Narlikar theory and also Dirac's cosmology with additive matter creation.

McElhinny's improved analysis of paleomagnetic data for continental blocks confirmed and sharpened his conviction that there had been no systematic change in the Earth's radius since the Devonian. At a confidence level of 95 % he concluded that the average paleoradius was $R_E/R_0 = 1.020 \pm 0.028$. It followed that the maximum possible expansion rate was 0.13 mm per year, "a value sufficiently small not only to exclude fast expansion rates proposed by Carey (1958, 1977) and Hilgenberg (1962), but also to exclude the much slower rates proposed by Egyed (1963) (1 mm per year) or Wesson (1973) (0.6 mm per year)."[109] To the mind of McElhinny and most other geophysicists, there could be no doubt that "all expanding earth hypotheses can confidently be rejected."

According to Peter Smith, a geophysicist at the Open University, the day that McElhinny and his collaborators published their paper was "a bad day for the handful of people who support the idea of an expanding Earth, [but] a good day for the few who oppose it."[110] Notice that "the few" referred to earth scientists actively opposing the expanding Earth hypothesis and not to those who simply ignored it. On the other hand, the result of McElhinny and co-workers concerning the gravitational constant was based on expansion rates for the Earth and therefore more indirect and less accurate than the one derived from astronomical distance methods. The same was the case with a study by the Dutch astronomer J. van Diggelen, who used Wells' coral growth method to conclude that there is "no evidence for any expansion of the Earth in the past 5×10^8 year."[111] He consequently suggested that Van Flandern's conclusion of a decreasing gravitational constant was erroneous.

Finally, G = constant was convincingly demonstrated by the Lunar Laser Ranging Project starting in the late 1960s and originating to a large extent in suggestions made by Dicke and his Princeton group even before the invention of the laser.[112] In July 1969 the first reflectors were placed on the Moon by astronauts from the U.S. Apollo 11 mission. Over the next couple of decades more than 10,000 range measurements were made. Analysis of data published in 2004 resulted in the limit

---

[107] Dicke (1962a). See Sect. 3.3.

[108] McElhinny et al. (1978), p. 320.

[109] McElhinny (1978), p. 152. For the reference to Hilgenberg, the veteran of Earth expansion, see Sect. 3.1 and Hilgenberg (1962). McElhinny's reference to Carey (1977) is erroneous and should possibly be to Carey (1976).

[110] Smith (1978).

[111] Van Diggelen (1976).

[112] Bender et al. (1973).

$$\frac{1}{G}\frac{dG}{dt} = (4 \pm 9) \times 10^{-13}\,\text{year}^{-1}.$$

Three years later a similar study concluded that

$$\frac{1}{G}\frac{dG}{dt} = (2 \pm 7) \times 10^{-13}\,\text{year}^{-1}.$$

The two studies seem to rule out all theoretical models of varying gravity.[113] For all that we know, the gravitational constant is in fact constant, in full agreement with the standard Einstein theory of general relativity. Of course, experiments can never prove G to be constant *precisely*, but only that the time variation is smaller than a certain quantity given by the experimental uncertainty. All the same, Dirac was wrong.

4.4   Degeneration

The new plate tectonics had no need for a decreasing G, a hypothesis that also seemed unnecessary to most astronomers and cosmologists. One might therefore believe that interest in the $G(t)$ hypothesis waned through the 1970s and 1980s, finally to disappear, but this is not what happened. Quite on the contrary, scientific interest in the hypothesis increased markedly, with a growing amount of publications on the subject written primarily by physicists and astronomers.[114] On the other hand, references to the $G(t)$ hypothesis decreased markedly in the geophysical literature. Noting that "we are in the middle of a variable-G research boom," in 1981 Wesson expressed some surprise concerning the popularity of the subject. After all, it was hard to justify research in the varying-gravity hypothesis from either a theoretical or an observational point of view (Fig. 4.3).[115]

The development up to the early twenty-first century demonstrates how studies of $G(t)$ continued after the acceptance of plate tectonics and the hot big bang, indeed after Dirac's death in 1984.[116] Whereas the story of the varying-G conjecture is one of an interesting mistake, the more recent story of the expanding Earth is different. It is a story of how a scientific hypothesis gradually degenerated into what can reasonably be described as a pseudo-science or something close to it.

By the early 1970s the expanding Earth was still considered a possibility by many earth scientists but mostly in the slow-expansion version of Egyed, Creer,

---

[113] Williams et al. (2004), Müller and Biskukep (2007).

[114] Reviews are given in Wesson (1978) and Gillies (1983). See also the citations to Dirac's (1938) paper in Fig. 2.3.

[115] Wesson and Goodson (1981).

[116] For a useful summary account of various modern constraints on $G(t)$, including arguments from astronomy, cosmology, geophysics and palaeontology, see Uzan (2003), pp. 25–31.

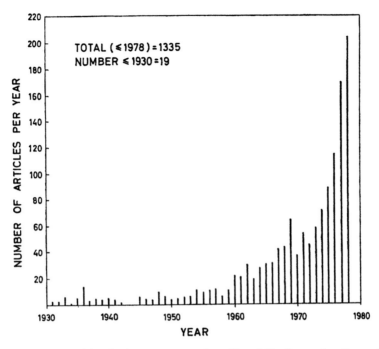

Fig. 4.3 Number of articles "having *some* connection with variable-*G* cosmology" according to Wesson and Goodson (1981) and based on the bibliographies in Wesson (1978) and Wesson (1980). The variation for 1937–1978 is in rough agreement with the one shown in Fig. 2.3, but the numbers undoubtedly exaggerate interest in varying-*G* theories. Notice the many publications before Dirac's first papers in the late 1930s. Reproduced with the permission of The Observatory Magazine

Dearnley and others. In a review of expansion hypotheses dating from 1973, James C. Dooley at the Bureau of Mineral Resources, Canberra, developed a test based on the geometrical consequences of Earth expansion.[117] Noting that "the chain of arguments leading to the conclusion that the earth is expanding is long and complex," he concluded that the rapid expansion proposed by Carey could not have occurred. As regards the slower expansion at the rate 0.5–0.6 mm per year he thought it was a possibility but not one supported by strong evidence. Dooley's cautious and somewhat sceptical attitude was shared by several other geologists who were interested in an expanding Earth without being committed to the hypothesis (see also Sect. 3.6).

As mentioned, at the time when Jordan's monograph on the expanding Earth appeared in an English translation (1971), scientific interest in expansion was in decline, a process that only accelerated through the decade. Still in 1966 Scheidegger admitted Earth expansion as a possibility, although without having

---

[117] Dooley (1973), who referred to the varying-*G* hypothesis entertained by Dirac, Jordan, Egyed and Dicke but without commenting on the validity of the hypothesis.

much confidence in it. Not only was it difficult to bring the hypothesis "into accord with the commonly accepted principles of physics," he also pointed out that expansion implied a homogeneous stress state over the Earth's surface in disagreement with observed horizontal displacements.[118] According to Scheidegger, among the available geodynamical theories continental drift combined with convection currents was the best offer. Ten years later, looking back at "the host of geotectonic hypotheses which had to be discarded at the end of the 1960s," Scheidegger mentioned "the time-honored contraction hypothesis" as the most important of the failed hypotheses. But he also called attention to "the opposite possibility, that of an expanding Earth, [which] also had to be discarded."[119]

The expansion theory or hypothesis was indeed discarded in the sense that the large majority of mainstream geophysicists considered it to be erroneous, preferring to ignore the expansion theory rather than arguing against it. And yet it was alive and defended by a minority of earth scientists. As Anthony Hallam noted with some surprise in an essay review of 1984, the number of adherents to Earth expansionism had not diminished. There were even "some very respectable geologists" who sympathized with the idea.[120] A few years later Le Grand similarly judged that expansionism "may have more adherents now than in the late 1960s."[121] Indeed, according to a bibliographic data base prepared by Giancarlo Scalera, a geophysicist and modern expansionist, the number of publications on or relating to the expanding Earth peaked in the 1970s and 1980s with 185 and 293 publications, respectively (Fig. 4.4). His list of 1040 sources "devoted to the expanding Earth" gives the following distribution over time:

1900–1919	7 publications
1920–1939	51 publications
1940–1959	86 publications
1960–1979	357 publications
1980–1999	533 publications

However, the list is grossly misleading as it includes a large number of publications which are in no ways devoted to or just relate to the expanding Earth hypothesis.[122] Whereas many advocates of the expanding earth were geologists, the hypothesis was also defended by amateurs, sometimes in unexpected places. For example, in 1971 an article on the subject and its relation to varying gravity

---

[118] Scheidegger (1966), p. 141.

[119] Scheidegger (1976), p. 143.

[120] Hallam (1984), who did not identify the respectable geologists.

[121] Le Grand (1988), p. 253. Moreover: "Expansionists are not ruled out as 'cranks' even by their opponents and their articles continue to be published" (p. 263). Ten years later, expansionists were, by and large, ruled out as cranks.

[122] Scalera and Jacob (2003), pp. 419–465. Strangely, Jordan's important *Schwerkraft und Weltall* is missing from the bibliography.

Increasing number of papers about expanding Earth

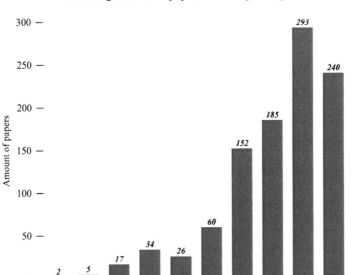

Fig. 4.4 Growth in the literature on the expanding Earth. Reproduced from Scalera and Jacob (2003) with the permission of Giancarlo Scalera

appeared in the *Journal of Naval Science* published by the UK Ministry of Defence.[123]

One of the late expansionists was Hugh Owen who continued to defend the hypothesis of an expanding Earth and develop it into versions based on speculations concerning the dynamics of the physical state of the inner Earth. As late as 2012 he confirmed his belief in an expanding Earth.[124] Another late expansionist was Johann Steiner, who had earlier defended the varying-*G* hypothesis. In a paper of 1977 Steiner attempted to breathe new life into the expanding Earth hypothesis, this time by relying on geological data alone and without referring to the cause of the expansion.[125] Steiner reasoned that for a constant Earth the rate of sea-floor spreading during a given period of time must be approximately equal to the rate of subduction in the same period. He estimated that since the beginning of the Jurassic, some 165 million years ago, the sea floor area had increased by 431 million km^2 while the increase in global subduction had only been 325 million km^2. The 33 % difference he ascribed to an increase in the Earth's radius, which in the Jurassic had been 5668 km or 89 % of its present radius. However, the estimated

---

[123] Kirby (1971). The author, who supported the expanding Earth theory, was a science officer employed by the Admiralty Underwater Weapons Establishment.

[124] Owen (2012), who pointed out the similarity between his ideas of the inner core of the Earth and those proposed by J. Halm in 1935, as mentioned in Sect. 1.3.

[125] Steiner (1977).

error in the paleoradius was ± 13 %, which made his argument less convincing. Steiner's analysis was welcomed by the dwindling expansionist camp but failed to convert anyone of the large majority of non-believers.[126]

In fact, Steiner's article was severely criticized by several mainstream geologists in the pages of *Geology*, the journal of the American Geological Society in which the article was published.[127] As one of the critics, William Kaula of the University of California, Los Angeles, pointed out, Steiner's conclusion lacked a physical mechanism. If the supposed expansion was due to a weakening G it would require a rate of about 2×10^{-9} per year, much greater than allowed by Van Flandern's recent measurements. Another of the critics, Norman Sleep of the North-Western University, Evanston, concluded that "The speculations presented by Steiner are invalid and result from the misuse of data tabulated for other purposes." But he also mentioned a more general reason to disbelieve in expansion: "The internal consistency of plate tectonics data and the elementary laws of physics remain the best arguments against significant changes in the Earth's radius." In reply to his critics Steiner maintained the validity of his conclusion. He had no respect for the appeal to internal consistency and the consensus view of the laws of physics. "A popularity poll is not necessarily the best way to obtain the truth," was his answer.

At the time of the Sydney symposium of 1981, the large majority of earth scientists considered the expanding Earth a lost case that was not worth further consideration. They were aware of the hypothesis but preferred to ignore it rather than criticizing it. Nonetheless, two years before the Sydney symposium the venerable Geological Society of London[128] arranged a meeting on the expanding Earth with speakers arguing for and against the hypothesis. The latter group was represented by Runcorn and Stewart who both argued against models of strong expansion such as Carey's.

The report of Robert Muir Wood in *New Scientist* gives a vivid but also partisan impression of the low reputation expansionism had at the time in some quarters of mainstream geophysics. As Wood described the meeting, during the course of "a theatrical afternoon" the hypothesis of an expanding Earth was defended by "a bombastic Tasmanian professor of geology (Warren Carey) and a less flamboyant English geophysicist (Dr Hugh Owen)."[129] Wood suggested that Carey's new idea of matter creation under high temperature and pressure was plainly unscientific, nothing less than a return to "medieval mysticism." After listing some arguments against the hypothesis of Earth expansion, he concluded his attack: "Such ideas as

---

[126] See Smith (1977).

[127] See *Geology* **6** (1978): 377–383, with Steiner's reply on pp. 379–383.

[128] Founded in 1807 the Geological Society is the oldest society of its kind and one of the oldest scientific societies devoted to a particular discipline. The oldest is the Linnean Society, which was founded in London in 1788 as a society for the study of botany and natural history generally.

[129] Wood (1979). See also Wood (1985), pp. 208. A less entertaining but more objective review of the meeting was given by the Cambridge geologist W. Brian Harland according to whom the expanding Earth was a challenging but probably wrong alternative to the standard view of a constant-sized Earth. See Harland (1979).

the expanding Earth with its biological metaphoric supply from growth and preg-
nancy, provide entertainment for a cold winter afternoon."

Understandably, Carey wrote a reply to Wood's insults, but *New Scientist*
decided not to publish it.[130] In a book published six years later, Wood repeated
his complete disrespect for the expanding Earth hypothesis, unfortunately offering a
caricatured picture by identifying expansionism with Carey's extreme version of
it. According to Wood, the idea of the expanding Earth was a return to an
old-fashioned geocentricism with roots in nineteenth-century geology based on
the presupposition that "the science of the Earth operates according to laws
different from those of the heavens."[131] This is however quite misleading and
especially so with regard to expansion theories based on a changing gravitational
constant. These theories were attempts to adopt a universal law of nature to the
Earth and in general to integrate elements of cosmology with elements of the earth
sciences. Contrary to Wood's claim, Earth expansionism was no more geocentric
than was plate tectonics. Besides, most nineteenth-century geology rested on the
uniformitarian principle which ruled out laws of nature that were valid only for the
Earth.

At any rate, from about the time of the London meeting, expansionism was
increasingly drifting apart from mainstream geophysics and indeed from mainstream science. In the terminology of the influential Hungarian-British philosopher
of science Imre Lakatos, it was a research programme that had entered its
degenerating phase.[132]

Although the work of McElhinny and collaborators may have been bad news for
supporters of the expanding Earth, it did not silence them. Without questioning the
validity of the geomagnetic evidence provided by McElhinny and others, in 1981
two Australian geophysicists, P. W. Schmidt and B. J. Embleton, argued that in the
Proterozoic the Earth might well have expanded significantly.[133] In agreement with
Creer's earlier assumption they suggested that the radius of the primitive Earth was
only 0.55 times the present radius. The two Australians ended their paper by
repeating Creer's warning of 1965, that it might not be legitimate "to extrapolate
the laws of physics as we know them to times of the order of the age of the Earth."
They did not speculate about the cause of the expansion.

As yet another late advocate of expansionism, consider the highly respected
South African geologist Lester Charles King, who in the 1930s and 1940s had been
one of the few supporters of continental drift and in 1956 participated in the Hobart
symposium organized by Carey. Unable to accept the new plate tectonics in general
and sea floor spreading in particular, he turned to expansionism. Arguing the case
purely empirically, he had nothing to say about the force responsible for the

---

[130] Carey (1988), p. 173.

[131] Wood (1985), p. 209.

[132] Oldroyd (1996), p. 275. According to Lakatos, a theory or research programme is degenerating
if it is kept alive only by auxiliary hypotheses that do not result in greater explanatory or predictive
power. A theory is pseudo-scientific if it fails to make novel predictions.

[133] Schmidt and Embleton (1981).

expansion. While sea floor spreading was not a problem for Carey and most other expansionists, King dismissed it scornfully. "The hypothesis of continuous seafloor spreading from Jurassic to recent is not tenable," 76-year-old King wrote in a book presenting his view of the expanding Earth.[134] Also contrary to Carey, King maintained that the Earth had expanded early on very rapidly and reached its present size before the Mesozoic era ended. While all the major oceans had grown in size throughout the Phanerozoic, the continents "have been essentially the same size through 2000 million years." King explained:

> The difference between oceans basins is apparent, and is measurable, as "continental drift." It is like the parade of soldiers who were moved individually (and surreptitiously) so that each was spaced from his fellows by twice the original distance. Each thought that the others had moved away from him; but an enlargement of the space they all occupied had brought about the antipathy evident in each case.[135]

The new plate tectonics based on continental drift and sea floor spreading was primarily a Western science dominated by researchers in the United States, Canada, Great Britain, and Australia. The theory was initially met with serious opposition by Russian geologists and geophysicists. Vladimir Beloussov, a prominent geologist at Moscow University, was an outspoken critic of plate tectonics and sea floor spreading not only in his own country but also abroad. In 1960 he was elected president of the International Union of Geodesy and Geophysics and four years later he became the Soviet editor of the international journal *Tectonophysics*. "Not a single aspect of the ocean-floor spreading hypothesis can stand up to criticism," Beloussov declared in an article of 1970, regretting that Western plate tectonics had evolved into what he called a self-confident dogma. Critical voices, he wrote, "can hardly be heard over the shouting of the fanatic adherents of the ocean-floor-spreading hypothesis."[136] One might perhaps expect that Beloussov, in his discussion of alternatives to the allegedly dogmatic theory of plate tectonics, appealed to the somewhat similar criticism raised by Carey and other Western expansionists. But this was not the case.

Although several Russian earth scientists favoured a kind of expansion or pulsation hypothesis, it was typically in the limited sense of tectonic processes increasing or oscillating in time. For example, Evgenii E. Milanovsky, another important Moscow geologist, wrote several papers on this kind of alternative to plate tectonics in *Tectonophysics*. He followed Beloussov in never referring to the Western tradition of expansionism. Among the few Russians who supported expansion in the sense of a steadily increasing Earth radius was Elena Lubimova (see Sect. 3.4). Even though there may not have been much interest among Russian scientists in the ideas of Carey, Egyed and Creer, there was an interest in expansion

---

[134] King (1983), p. 142. According to Hallam (1984), the book was "unlikely to produce any converts to an expanding Earth model."

[135] King (1983), p. 176.

[136] Beloussov (1970), p. 506. On the hostile reception of continental drift and plate tectonics in the USSR, see Khain and Ryabukhin (2002) and the more contextual account in Wood (1985), pp. 210–223.

or pulsation theories which were the subject of a large conference that Milanovsky organized in Moscow in 1981.[137] According to Victor Khain and Anatoly Ryabukhin, still at the turn of the century, "the hypothesis of an expanding Earth is rather popular among some Russian geologists."[138]

The hypothesis of Earth expansion driven by decreasing gravity seems to have attracted only limited attention among Russian physicists interested in gravitation and general relativity theory. In a paper of 1966 the distinguished theoretical physicist Dmitri Ivanenko and his co-workers Boris Frolov and V. S. Brezhnev investigated Dirac's $G(t)$ hypothesis and its relation to Einstein's cosmological field equations. Applying the hypothesis to the Earth, "we obtain for the rate of the earth's expansion $[dR/dt] \cong 0.05$ mm/year, which curiously coincides with the values on the basis of geological data." They noticed that the expanding Earth hypothesis was "still not generally accepted."[139] Four years earlier Ivanenko and Frolov had organized a symposium in Moscow on "Related Problems in Gravitation and Geology."

Cosmology and geology are sciences rich in controversies and the one here considered is only exceptional because it was neither purely cosmological nor purely geological. Historians and sociologists have dealt extensively with the concept of scientific controversy.[140] It is generally agreed that in order for a scientific disagreement to qualify as a controversy it should be of some duration, be expressed in public, and take place by means of arguments and counterarguments. Moreover, a controversy is more than just a debate or a discussion: the parties must be committed to one of the opposing views and attack the rival view. Only if the relevant scientific community considers the disagreement worth taking seriously will it evolve into a controversy. On this view there is no doubt that the discussion concerning the expanding Earth was a real controversy although it was far less important than the greater one between the fixed Earth picture and the theory of continental drift.

Some controversies are about facts, meaning that scientists disagree about the empirical basis of a knowledge claim, that is, whether the claimed property or phenomenon exists or not. A controversy of theory involves different theoretical

---

[137] See Carey (1988), p. 137, according to whom the conference was attended by no less than 700 participants and resulted in 20 papers.

[138] Khain and Ryabukhin (2002), p. 194, which is an extended version of the earlier review Khain (1991).

[139] Brezhnev et al. (1966). Ivanenko was an old acquaintance of Dirac, whom he had first met in 1928. He did important work in quantum mechanics and nuclear physics and later in general relativity and unified field theory. A member of the International Committee on Gravitation and General Relativity, Ivanenko was instrumental in organizing Soviet research in gravitation physics and establishing in 1962 a Soviet Gravitation Commission. In 1961 Ivanenko wrote jointly with M. U. Sagitov a paper titled "The secular increase [sic] of gravity and the expansion of the Earth." I have not seen this paper, which seems to have been published in Russian only. The source is given in Carey (1975), p. 142.

[140] See, e.g., Engelhardt and Caplan (1987) and Machamer et al. (2000).

views, whereas a controversy of principle relates to, for example, basic methodological or ontological principles. The three categories are not mutually exclusive, of
course. The controversy about the expanding Earth, and also the one concerned
with varying gravity, was primarily a controversy of fact with an element of
controversy of theory. While many scientific and technical controversies involve
consequences of a social, religious or political nature, this was not the case at all
with respect to the expanding Earth and varying gravity.

Scientific controversies end in different ways.[141] A controversy may be resolved
if the participants and the relevant scientific communities agree that one of the
views is correct and the contesting view is incorrect; or a controversy may come to
an end by the intervention of external authority even though the original disagreement still persists to some extent. Finally a controversy may just wither away, be
abandoned because the scientists lose interest in it. The case of the expanding Earth
contains elements from the first and last of the categories in particular although the
controversy cannot be said to have been finally resolved (given that there are still a
few scientists defending the expansionist cause). At any rate, supporters of an
expanding Earth hypothesis were marginalized and the community of earth scientists tacitly decided that it was no longer worthwhile to deal with the hypothesis.

Textbooks in global geophysics from the 1980s would typically leave Earth
expansion unmentioned, just as textbooks from the 1940s typically left continental
drift unmentioned; or as textbooks in astrophysics and cosmology from the 1970s
typically ignored the steady-state universe. The effect was that many young geophysicists would not even know what expansion was all about.

The way the controversy over the expanding Earth reached closure has some
similarity to the contemporaneous cosmological controversy between relativistic
evolution theory and the steady-state theory of the universe. However, there were
also dissimilarities. In the cosmological case the chance discovery of the cosmic
microwave background radiation was of decisive importance, whereas there was no
corresponding "smoking gun" event in the case concerning the Earth. No new
insight, experiment or observation made it suddenly obvious that plate tectonics
was the correct answer, while the expanding Earth was just a wrong idea. Another
difference was that the cosmological controversy included elements of a philosophical and religious nature. Whether justifiable or not, the big bang was sometimes
associated with theism and the steady-state universe with atheism. The religious
element was wholly absent from the debate over the Earth. Although methodological discussions entered at some occasions, discussions of a wider philosophical or
ideological nature played almost no role.

In spite of the de facto exclusion of the expanding Earth theory from mainstream
science, supporters of the theory continued to explore it and argue for its advantages
relative to the now dominant theory of plate tectonics. Carey was the undisputed

---

[141] See Engelhardt and Caplan (1987), which includes an article by Henry Frankel on the
continental drift debate (pp. 203–248). On the end of the cosmological controversy involving
the steady-state theory, see Kragh (1996), pp. 389–395.

leader of attempts in the 1970s and 1980s to fight the new orthodoxy—he was a kind of equivalent to Hoyle in relation to the big-bang theory. Also the German engineer Klaus Vogel should be mentioned. While still a citizen of the German Democratic Republic Vogel made elaborate models of the continents fitting on a globe nearly half the size of the equivalent Earth. By placing the primitive globe inside a transparent globe of the present Earth he offered a visual argument for radial expansion as the cause of the separation of the continents.[142]

To make a long story short, the non-existence of an appreciable expansion of the Earth has recently been established much more precisely and directly than earlier. Using advanced geodetic methods and measurement techniques, in 2011 an international team of researchers led by Xiaoping Wu from NASA's Jet Propulsion Laboratory published a study that "provides an independent confirmation that the solid Earth is not getting larger at present, within current measurement uncertainties."[143] Although Wu and his team of American, French and Dutch scientists found a tiny expansion rate of about 0.1 mm per year, with a measurement uncertainty of $\Delta R/\Delta t = 0.2$ mm per year the expansion was not statistically significant.

Nonetheless, this recent result seems to have had no effect on the believers in Earth expansion. For one thing, although expansion may not occur presently, it does not rule out that it occurred only in the geological past, such as claimed by King. Modern expansionists tend either to ignore Wu's result or suggest that it does not follow from the measurements because of a poor sensitivity-signal ratio. For example, the measurements are claimed to be too insensitive to rule out a radial expansion of the Earth of about 1.5 mm per year.[144] The price to be paid for this particular hypothesis is that the Moon must *contract* radially at the implausibly high rate of 41 cm per year.

I shall deal only briefly with the more recent attempts to keep the expanding Earth alive, with or without the varying-G hypothesis.[145] The main protagonists of modern expansionism are the Italian geophysicist Giancarlo Scalera and the Australian geologist James Maxlow who agree that continental drift is just an illusion created by the insistence that the radius of the Earth is constant. Whereas the most credible expansion Earth models in the 1960s and 1970s were the slow-expansion models of the type suggested by Egyed and Dicke, the more recent expansionists are in favour of rapid expansion—the very kind of expansion which repeatedly has been shown to disagree with measurements. Carey's stubborn

---

[142] Carey (1988), pp. 266–269. See Vogel (1992) for a summary of his arguments for the expanding Earth hypothesis.

[143] Wu et al. (2011). According to NASA, the team "confirmed what Walt Disney told us all along: Earth really is a small world, after all." http://www.nasa.gov/topics/earth/features/earth20110816.html

[144] Nyambuya (2014). For another recent attempt to show that the Earth is presently expanding at a slow rate (about 0.2 mm per year), see Shen et al. (2011).

[145] These attempts are critically reviewed in Sudiro (2014), where further references to the literature can be looked up.

refusal to accept any subduction is widely followed. In the 1990s Scalera argued that the Archaean radius of the Earth had been 3500 km, to increase in the Paleozoic to 4300 km and in the Mesozoic to 5300 km.

Some of Maxlow's ideas are even more extreme. He has proposed a primeval Earth with a radius as small as 1700 km, or the same as the Moon. Its density was as high as 290 g cm^{-3} and its surface gravity no less than $g = 138$ m s^{-2}. Other modern expansionists have suggested that the original mass was much less than today and that the Earth's mass has increased at a rate about 10^{15} or 10^{16} kg per year. Only few modern expansionists have appealed to a decreasing G as the mechanism behind the expansion. The Dirac–Jordan–Dicke hypothesis is still of some interest to physicists, but it seems to play almost no role at all in what is left of Earth expansionism.

As seen from the perspective of mainstream geophysics the theory of the expanding Earth has long been dead. Still in the late 1990s the historian and philosopher of science Richard Nunan concluded that as far as the empirical validity of expanding Earth models is concerned, "the jury is still out." Moreover, "The fortunes of either moderate or fast expansion could improve during the coming decades."[146] This view was definitely not shared by the large majority of geophysicists at the time. Today we know that the fortunes of expansion have not improved and we have good reasons to expect that they never will improve. Despite its poor scientific reputation, papers, symposia and conferences on the expanding Earth continue to this day. Paolo Sudiro summarizes the present state of Earth expansionism as follows:

> While expansionists claim that Earth scientists dogmatically follow a theory (plate tectonics) falsified by geological data, they promote or incorporate borderline and pseudoscientific ideas, including generation of new matter inside Earth, variation of cosmic constants, and exotic matter transformations, conflicting with accepted physical theories.[147]

This is not, in my view, an unfair characterization. A perusal of the extensive literature on current Earth expansionism in the form of books, papers and websites confirms the suspicion that the idea of Earth expansion, which for a period of one or two decades was a seriously discussed scientific hypothesis, has now deteriorated into something dangerously close to a pseudo-science.[148] The modern theory is primarily of interest from the point of view of sociology and psychology of science.

---

[146] Nunan (1998), p. 249.

[147] Sudiro (2014), p. 144.

[148] See, for example, the two symposia proceedings Scalera and Jacob (2003) and Scalera et al. (2012). An impression of the style of science in current Earth expansionism may be gained from Maxlow's website http://www.bibliotecapleyades.net/ciencia/earthexpanding/00_GlobalExpansionTectonics.htm#menu.

4.5   Two Revolutions in Science

Plate tectonics was not the only scientific revolution that occurred in the mid-1960s. Cosmology experienced one as well. As global plate tectonics was in part a revival of Wegener's old theory of continental drift, so the big-bang theory was in part a revival of the older explosion theories of the universe going back to Lemaître in the early 1930s and to Gamow and his collaborators Alpher and Herman in the late 1940s.

Although most physicists and astronomers in the first two decades after World War II favoured an evolutionary universe governed by the laws of general relativity, finite-age models with an explosive beginning at the origin of time enjoyed little support. A significant minority thought that the steady-state theory proposed in 1948 by Fred Hoyle, Hermann Bondi and Thomas Gold offered a better and more appealing picture of the universe.[149] First and foremost, at the time there was no shared foundation of cosmology, a science that was widely considered semi-philosophical because of its lack of relevant observations to decide between rival models. Preference of one cosmological model over other models was widely seen as a matter of taste, not something that could be justified in strict scientific terms.

The unclarified situation only began to change around 1960, when radio-astronomical measurements proved irreconcilable with predictions from the steady-state theory. The data clearly favoured relativistic evolution cosmology, if not necessarily of the explosive type. By the early 1960s radio-astronomical measurements had seriously weakened confidence in the steady-state theory, but not to the extent that it could be excluded as a possible alternative to the models described by general relativity. The turning point came in the spring of 1965 with the recognition that the newly discovered microwave background at wavelength 7.3 cm could be interpreted as a fossil radiation from the early phase of the big-bang universe.[150] While the cosmic microwave background fitted beautifully into the big-bang theory, it came as an unpleasant surprise to the supporters of the steady-state theory of the universe.

Preoccupied with the $G(t)$ hypothesis and its geophysical consequences—not to mention his work as a politician and author of popular works on science and culture—Jordan was not much involved in the cosmological debate that led to the standard hot big-bang model. He was aware of the steady-state model at an early time, but apart from finding the element of matter creation to agree with his own theory he did not support it. Instead he chose to emphasize "the considerable differences between Hoyle's theory and my own."[151] As he pointed out in

---

[149] For a comprehensive account of the steady-state theory and its relation to relativistic evolution theories, see Kragh (1996).

[150] Peebles et al. (2009) is a detailed account of the early history of the cosmic microwave background. As noted in Sect. 4.2, the discovery caused serious problems for Dirac's $G(t)$ cosmology.

[151] Jordan (1949), p. 640, Kragh (1996), p. 196.

Schwerkraft und Weltall, the steady-state theory did not recognize the unidirec-
tional nature of cosmic evolution and it also denied or disregarded the $G(t)$
hypothesis which Jordan valued so highly.[152] In his 1966 monograph on the
expanding Earth, Jordan referred to the steady-state theory and its recent refutation
by radio astronomers' discovery of "what must be interpreted as the remnant
radiation of the big bang."[153] As mentioned in Sect. 2.6, a couple of years later
the microwave blackbody background forced him to modify his own cosmological
theory.

Contrary to Jordan, Dicke and his group in Princeton were crucially involved in
the radical transformation that cosmology experienced in the 1960s. As we have
seen, Dicke was strongly attracted to the scalar–tensor theory with a decreasing
gravitational constant. This theory guided much of his thinking about the early
universe which in 1963 led him to consider an initial hot phase filled with blackbody radiation. To a large extent, what became the standard big-bang theory was
indebted to a heterodox theory of gravitation, namely the Brans–Dicke theory. It
was also indebted to another heterodox view, namely, Dicke's preference for a
cyclic universe and his belief that the big bang had its origin in the contraction of a
preceding universe into a big crunch. As a result of the contraction, radiation in the
earlier universe would be strongly shifted to the high-energy blue frequency region
and therefore capable of destroying heavy atomic nuclei by means of photo-
dissociation. This belief was visible in the paper that Dicke and his three Princeton
co-authors published in 1965, but it was of no importance to the argument of the
paper.

The significance of the Brans–Dicke theory is evident from a review paper that
Dicke and Peebles submitted in early March 1965, shortly before they became
aware of the discovery of the 7.3 cm background constructed by Arno Penzias and
Robert Wilson the previous year.[154] In this paper the two Princeton physicists dealt
at length with the geophysical, astrophysical and cosmological consequences of the
Brans–Dicke theory. Using the standard equations of general relativity, at the time
Peebles had reached the conclusion that the intensity of the present background
radiation would correspond to a temperature of about 10 K to avoid excess helium
production in the past. Should the temperature of the radiation turn out to be
considerably less, Dicke and Peebles appealed to the decreasing gravitational
constant of the Brans–Dicke theory. Because, with a larger G in the cosmic past,
"the universe would have expanded through the early phase very much faster than is

---

[152] Jordan (1955), pp. 136–138.

[153] Jordan (1966), p. 138. Preface dated February 1966. While the English translation used
"primordial 'big bang'," the German original referred to the *Urknall*, a word for the explosive
event in the past which at the time was common in German literature and is still used.

[154] Dicke and Peebles (1965). In a note added in proof Dicke and Peebles referred to the Penzias-
Wilson discovery and their own interpretation of the radiation as a fossil from the early hot
universe.

implied by general relativity [and] this would reduce the time available for helium production, thus reducing the lower limit on the present radiation temperature."[155]

A passage to the same effect appeared in the seminal paper in the July 1965 issue of the *Astrophysical Journal* in which Dicke and his co-authors Peebles, Peter Roll and David Wilkinson analysed and interpreted the cosmic microwave background.[156] A few years later Dicke returned to the problem of helium production in scalar–tensor cosmology, where the scalar field contributes significantly to the expansion rate. His result was not encouraging: "For a flat universe (present density 2×10^{-29} g cm^{-3}, Hubble age 10^{10} years), zero helium production would be expected. For the low-density case (present density $\sim 7 \times 10^{-31}$ g cm^{-3}) there are two possibilities: 32 per cent and 0 per cent helium."[157] From a modern perspective it may appear that considerations of varying gravity in the formation of the new big-bang theory were really irrelevant and unnecessary. However, they actually played a considerable role and indirectly provided a link to the kind of geophysical work that was similarly motivated. As Dicke saw it, both the Earth and the universe might serve as a testing ground for the scalar–tensor theory of gravitation that was the primary target for his long-term research project.

Plate tectonics and the hot big-bang universe appeared at about the same time. They were turning points in the earth sciences and cosmology, respectively, and they have both been described as scientific revolutions. But were they really revolutions and, if so, were there any connections between these two revolutionary phases in the recent history of science?

With regard to the last question I have argued that there was indeed such a connection, namely the one mediated by the $G(t)$ hypothesis; but I have also suggested that this connection was indirect and not, after all, very important.[158] The first question is somewhat peripheral to the subject of this study, for which reason I shall only deal with it cursorily. In retrospect it is tempting to see the change to the hot big bang in the late 1960s as a genuine revolution in our conception of the universe. This is today a common view, espoused not only by physicists and astronomers but also by some historians and philosophers of science. But from a contemporaneous perspective the change was of a different nature. After all, a finite-age universe governed by the laws of general relativity was accepted by a majority of cosmologists several years before the celebrated discovery of the cosmic microwave background.

---

[155] Dicke and Peebles (1965), p. 451, Peebles et al. (2009), pp. 38–39.

[156] Dicke et al. (1965), p. 418.

[157] Dicke (1968), p. 22.

[158] Theoretical physicist Werner Israel (1996) investigated in an interesting essay the parallel histories of continental drift and ideas of super-dense celestial bodies (such as neutron stars, quasars and black holes). Both concepts became part of established science in the 1960s. Whereas there are noteworthy methodological and sociological similarities between the two cases analysed by Israel, they are strictly separate with regard to substance, which is contrary to the cosmology-geophysics case.

As far as I know, none of the key actors in the transformation of cosmology, including pioneers such as Dicke, Peebles and the Russian astrophysicist Yakov Zel'dovich, described it *at the time* as a revolution. This is not to say that the term "revolution" cannot be found in the literature. For example, in 1967 Peebles and Wilkinson referred to the identification of the microwave background as "a revolutionary development in cosmology."[159] They did not elaborate or relate the recent cosmological revolution to, say, the Copernican revolution in the late renaissance or other revolutionary changes in the history of science. Peebles and Wilkinson also did not refer to Thomas Kuhn's view of scientific revolutions and paradigm shifts as relevant to the change their science had experienced. At the time they may have been unaware of Kuhn's famous book on *The Structure of Scientific Revolutions* published in 1962. If one looks for a revolution in twentieth-century cosmology, the shift at about 1930 from a static to a dynamic universe is a much better candidate than the shift in the mid-1960s.[160]

The well-documented case of continental drift and plate tectonics is quite different. As early as 1963, at a symposium on continental drift in Berkeley, J. Tuzo Wilson stated that the earth sciences were "ripe for a major scientific revolution." He presented the situation as corresponding to that of earlier pre-revolutionary phases in science, "like that of physics before quantum mechanics." Wilson did not mention Kuhn's *Structure* at this point but later recalled that he most likely had been inspired by Kuhn.[161] Five years later, in a paper read to the American Philosophical Society on 19 April 1968, he reflected on the recent developments in the earth sciences and how they related to Kuhn's "brilliant analysis of scientific methods." According to Wilson:

> Kuhn makes the point that a change in belief has been the essence of the great scientific revolutions like those from phlogistics [*sic*] to modern chemistry, from caloric to modern thermodynamics, or from special creation to evolution. In our day it would appear that what earth science needs more than fresh data, better instrumentation, or new techniques is a simple change from our present belief that the structure of the earth is static to the new concept that it has long been mobile. This is parallel and similar to the Copernican revolution and should perhaps be called the Wegenerian revolution from its chief advocate. ... We should start afresh and combine all investigations into the study of a mobile earth-structure under the name of one new science—geonomy.[162]

The last word in the quotation—"geonomy" as a terrestrial counterpart to "astronomy"—is not well known and in need of a comment. It was not Wilson's invention. Since the mid-nineteenth century the term was occasionally used in somewhat different meanings, often referring to the physical laws relating to the Earth. James Stanley Grimes, a nineteenth-century American geological author who used the term in the title of a book, defined geonomy as "a science which

---

[159] Peebles and Wilkinson (1967), p. 28.

[160] On this issue, see Kragh (2014b) and sources mentioned therein.

[161] Wilson's 1963 address is quoted in Cohen (1985), p. 464, and his later recollection on p. 565.

[162] Wilson (1968), p. 317. See also Wilson (1977).

relates to the physical laws of the earth, and includes all the essential facts of geology and physical geography."[163]

Wilson may have been the first geophysicist to describe the change in the 1960s as a revolution in something like Kuhn's sense, but he was not the only one. Within a few years he was followed by Anthony Hallam, Allan Cox, Ursula Marvin and half a dozen other geological authors.[164] Plate tectonics and revolution rhetoric went hand in hand. To the mind of Hallam, an Oxford geologist, there was no doubt that plate tectonics was the new paradigm of a recently finished revolution which finally had turned geology into a mature science. "The Earth sciences do indeed appear to have undergone a revolution in the Kuhnian sense," he emphasized in his triumphalist book on plate tectonics significantly titled *A Revolution in the Earth Sciences*.[165] Allan Cox, a leading actor in the construction of plate tectonics and a critic of the expanding Earth, affirmed that "the development of plate tectonics ... fits the pattern of Kuhn's scientific revolutions surprisingly well."[166] Echoing Hallam and Cox, the prominent planetary geologist Marvin declared that "the story of continental drift as a geologic concept ... bears out in a dramatic fashion a thesis developed by Thomas S. Kuhn."[167]

Among many working earth scientists there was a general consciousness that they lived during a revolutionary phase in their science, as witnessed by a series of early historical or semi-historical reviews by the scientists themselves. One looks in vain for a similar literature written by the physicists and astronomers who at the same time could celebrate the new big-bang theory of the universe. In most of the reviews written by geologists and geophysicists, the expanding Earth and varying-G theories were either ignored or just briefly mentioned. Yet also Carey referred to plate tectonics as a Kuhnian revolution, but understandably with a focus on the dogmatism that Kuhn associated with the adoption of a new paradigm. "When a new fact appears, it is automatically interpreted in terms of this ruling dogma, even though it may be equally or better explained otherwise," he complained. "Indeed, if it were explained in terms of Earth expansion, the report would certainly be sent back by journal referees for rewriting, if not rejected outright as naïve."[168]

Historians and philosophers of science are not so sure that global plate tectonics qualifies as a *bona fide* scientific revolution. Rachel Laudan concludes that although the emergence of plate tectonics can be described as a revolutionary change, it was

---

[163] Grimes (1858), p. 2.

[164] See Laudan (1980), Laudan (1983) and Cohen (1985), pp. 446–466. See also McKenzie (1977) for a comparison of the revolution in plate tectonics and the one of DNA-based molecular biology. According to McKenzie (p. 121), "plate tectonics was less fundamental a revolution than the discoveries which began molecular biology." A good overview is provided in Tetsuji Iseda, "Philosophical interpretations of the plate tectonics revolution," http://tiseda.sakura.ne.jp/works/Plate_tectonics.html

[165] Hallam (1973), p. 108.

[166] Cox (1973), p. 5.

[167] Marvin (1973), p. 189.

[168] Carey (1988), p. 197.

not a *Kuhnian* revolution. Among other things, there was no established paradigm or "normal science" in the previous decades, and there also was no incommensurability between pre- and post-tectonic geological theories. Nor was there any communication gap between the different camps.[169] The empirical-inductive methodology of geophysics before plate tectonics remained basically the same after the alleged revolution. The relative ease with which Tuzo Wilson moved from contractionism over expansionism to plate tectonics is but one example among many that illustrates the continuity of the development and non-Kuhnian nature of the revolutionary change.

4.6 Historiographical and Other Perspectives

The broader framework of this book is the historical relationship between cosmology and geophysics or between cosmology and the earth sciences generally. It is a topic that has so far attracted very little interest and of which there is no systematic study in the history of science.[170] More specifically I have presented a comprehensive account of two interrelated concepts, the one belonging to physical cosmology and the other to geology. The two concepts are the hypothesis of decreasing gravity and the hypothesis of the expanding Earth, and I have focused on how they interacted in the period from the late 1940s to the early 1980s. It is argued that it was only with Dirac's $G(t)$ hypothesis, or rather with Teller's critique of it from 1948, that modern post-Einsteinian cosmology became of importance to the earth sciences. Until then the only major problem area shared by the two sciences was the age of the universe and its relation to the age of the Earth.

Today it is generally agreed that the gravitational constant G does not vary or, to put it more cautiously, if it varies then the rate of variation is much slower than the one proposed by Dirac. Moreover, it is agreed that the Earth does not expand measurably. In other words, most of this lengthy study is about two mistaken hypotheses that for long have been abandoned by mainstream science and even in the 1960s were never more than minority views. The study may thus be of little or no relevance to current science and seem to be just so much ado about nothing. On the other hand, the case under investigation is not without interest from the point of view of history of science, a branch of scholarship that is not primarily concerned with current knowledge. In addition the case provides some insight into the problems that may occur when scientists from two very different disciplinary traditions face the same subject matter, *in casu* the Earth.

Much of the work reviewed in the book was of an interdisciplinary nature, in the sense that it dealt with subjects belonging to widely different fields of science. The

---

[169] Laudan (1980). See also Le Grand (1988), pp. 267–273 and Marx and Bornmann (2013).

[170] The title of Schröder and Treder (2007) is inviting but unfortunately misleading. The paper is not really a historical review of the connections between geophysics and cosmology.

expanding Earth and varying gravity appealed not only to geophysicists, astronomers and physicists, but also to palaeontologists, geologists and meteorologists— even a few biologists and chemists got interested. It is however characteristic of the period's literature that the interdisciplinary nature of the problem area was not reflected in a corresponding interdisciplinary authorship. Only in very few cases did physicists collaborate with geologists, or cosmologists with specialists in, say, paleomagnetism. Almost all papers were written by a single author or by a couple of authors with the same disciplinary background. As a consequence the authors were sometimes forced to deal with subjects outside their professional competences, which occasionally resulted in misunderstandings, poor judgments and superficiality. The way that Teller calculated the past climate of the Earth is one example; and the way that Egyed and Holmes received Gilbert's claim of the consistency of general relativity and varying gravity is another example. The element of amateurism that one can find in several of the publications is well illustrated in Jordan's extensive work and also in Hoyle's much briefer encounter with geophysics.

Jordan's work in geophysics was not only characterized by a certain degree of amateurism but also by some disrespect for the earth sciences when compared to fundamental physics. He found the methods of geophysics to be much less satisfactory than those of the fundamental branches of physics that he knew so intimately and where he felt better at home. One can find a similar attitude among some other physicists in the period, as when Dicke and Dyson commented on geophysics' deplorable lack of exactitude (Sect. 3.3).[171] Yet another example is Lyttleton, who was greatly interested in geophysics but whose attitude to the subject was very much astronomical rather than geological. His astronomy-inspired approach contributed to his dislike of continental drift and plate tectonics.

In 1953 the distinguished physicist and Nobel Prize winner Edward Appleton contributed to a festschrift for Max Born, his colleague at the University of Edinburgh. The subject of his paper was geomagnetism and the physics of the ionosphere. Appleton felt that he had to make excuses for the subject and its "general untidiness" which he contrasted to "the elegance which characterises Professor Born's own contributions to physical knowledge." As he pointed out, progress in these untidy geophysical fields "depends on the harvesting of sample measurements from different parts of our globe and attempting to reconcile them by tentative hypotheses and theories."[172]

It is noteworthy that the originator of the cosmological $G(t)$ hypothesis and its earliest and most articulate advocate were both pioneers of quantum mechanics, a theory Dirac and Jordan co-created while in their early twenties. Pascual Jordan

---

[171] According to Stephen Brush, fields such as geophysics and planetary science were for a long time considered "impure" and for this reason assigned an inferior status in the internal hierarchy of physics. See Brush (1996b), pp. 41–46.

[172] Appleton (1953), p. 1. At the time Vice-Chancellor of the University of Edinburgh, Appleton received the 1947 Nobel Prize in physics for his work on the physical properties of the upper atmosphere.

eventually turned towards the cosmology-geophysics connection, which he culti-vated for nearly two decades. The essay examines in some detail and for the first time how Jordan strove to vindicate Dirac's hypothesis by means of geophysical evidence. The remarkable role played by Joel Fisher, until now a persona incognita in the history of science, is described as far as the scanty sources allow.

Coming to the earth sciences as an amateur and with little respect for the culture of geophysics, Jordan was not highly regarded or even known among earth scien-tists. His extensive work on expansion of the Earth caused by weakening gravity did not attract a great deal of interest. Nor has it attracted any interest among modern historians of science.[173] The silence stands in stark contrast to the interest historians have shown to Jordan's seminal contributions to quantum theory and also, if to a lesser extent, his premature attempt in the 1930s to create a theory of quantum biology.[174] Jordan's extensive work in cosmology and geophysics has effectively been ignored.

Much of Jordan's work was independently duplicated and deepened by another prominent physicist and cosmologist, Robert Dicke in Princeton, who was more successful in crossing the barrier to the community of the geophysicists. What makes Dicke of particular interest is his seminal role in the revival of big-bang cosmology in 1965 following the discovery of the cosmic microwave background. As I point out, there was an indirect connection between the transformation in cosmology in the 1960s and the geophysical theories examined by Dicke on the basis of the Brans–Dicke $G(t)$ hypothesis. However, the connection was not very important, meaning that the Princeton physicists would most likely have arrived at their interpretation of the microwave radiation even if Dicke had not been preoc-cupied by the terrestrial and astronomical effects of a weakening gravitational constant.

I also speculate if the two scientific revolutions of the 1960s, namely global plate tectonics and the hot big-bang theory of the universe were somehow related. Given that the main mediator between cosmology and geophysics was the $G(t)$ hypothesis, and that this hypothesis played a role only with respect to the expanding Earth, it is more reasonable to see the two revolutions as developing in parallel but largely independently. The few connections between cosmology and geophysics one can find in the 1970s were non-mainstream in both areas, as illustrated by the theories of Hoyle, Wesson and Carey.

---

[173] Among the very few papers on Jordan's "excursion" into geophysics, see Kundt (2007) written by a former student and collaborator of Jordan. The article on Jordan in the authoritative *Dictionary of Scientific Biography* (vol. 17, supplement II) only mentions his work in the earth sciences with a single line, stating that "Jordan suggested explaining Wegener's continental drift phenomenon as a result of an expansion of the earth." However, Jordan considered the expanding Earth hypothesis as an *alternative* to continental drift and plate tectonics, which theories he did not accept. The updated entry on Jordan in the more recent *New Dictionary of Scientific Biography* (vol. 4) is no more informative with regard to his work in geophysics.

[174] On Jordan's work in quantum biology and his philosophical and ideological views, see Beyler (1994, 1996, 2009).

As far as the expanding Earth is concerned, the book mostly deals with the issue in relation to the varying-G hypothesis. It has not been my intention to write a full history of the idea of an expanding Earth, valuable as such a study might be. All the same, this book covers a substantial part of the history, from Jordan in the early 1950s to Carey and Owen three decades later. It only deals rather briefly with the later history. Today it is often forgotten that the epic debate about the Earth in the years from about 1955 to 1970 involved three and not merely two rival theories. Expansionism was evidently on the side of mobilism and against fixism, but at the same time it was opposed to continental drift in Wegener's sense of moving land masses. Although the debate concerning the expanding Earth is interesting in its own right, it should be kept in mind that the expansion hypothesis was relatively unimportant in comparison to the much better known main controversy that took place between fixists and drifters.

Moreover, varying gravity was only one aspect of many in relation to an expanding Earth, a hypothesis that in its modern version goes back to the mid-1950s but can be found decades earlier. There was little unity among the expansionists, some of whom focused on the mechanism driving the supposed expansion while others were unconcerned with its cause. Most were in favour of slow expansion, but an important minority (including Carey, Heezen, and to some extent Owen) argued for a substantially greater expansion rate. There were also considerable differences with regard to distribution of the expansion phases over geological time. According to a few expansionists (such as King) expansion no longer occurred, while others saw expansion as essentially limited to the more recent geological periods. Table 4.2 at the end of this section lists most of the expansion rates proposed in the period covered by the book, starting with Halm in 1935 and ending with Wu and collaborators in 2011.

It was generally accepted that decreasing gravity alone was insufficient to provide the necessary inflation of the Earth but there was no agreement as to its importance. Several earth scientists in favour of expansion disregarded the $G(t)$ hypothesis and some argued against it. Again, whereas some expansionists considered the expansion hypothesis an alternative to continental drift, there were also those who considered the two hypotheses to be compatible and complementary. Expansionists further differed considerably in their commitment to the hypothesis. Jordan and Carey were devoted to it to an almost dogmatic extent, whereas for others (Dicke, Wilson, and Holmes, for example) it was merely an interesting possibility that deserved to be examined. The lack of unity left expansionism in a much weaker position relative to the new plate tectonics than what otherwise could have been the case.

Although there was much opposition against the theory of plate tectonics in the 1970s and 1980s, there was no united front. Scientists might be sceptics or opponents for a variety of reasons and their positions did not necessarily imply a preference for a definite alternative such as an expanding Earth. It was customary to associate continental drift and plate tectonics, to see the first as a result of the latter, but some geologists denied the connection. While accepting the basic plate tectonic processes (such as sea floor spreading, subduction and transform faulting),

Table 4.2 Some Earth expansion rates

Scientist	Year	dR/dt	G(t)?	Comment
J. Halm	1935	1.6	No	Average expansion rate
B. C. Heezen	1960	~7	No	No definite rate given
J. T. Wilson	1960	~0.8	(Yes)	Finds G(t) to be "an inviting idea"
R. H. Dicke	1962	~0.05	Yes	Based on Brans–Dicke theory
D. van Hilten	1963	~4–7	No	No definite rate given
K. Creer	1965	0.75	No	Expansion starts 3.5 Gy ago
R. Dearnley	1965	0.65	(Yes)	Expansion throughout Earth history
A. Holmes	1965	~0.5	(Yes)	No conflict between expansion and continental drift
P. Jordan	1966	~5	Yes	Slow rate up to end of Palaeozoic
V. Brezhnev, D. Ivanenko, B. Frolov	1966	~0.05	Yes	Theoretical derivation based on general relativity
L. Egyed	1967	0.5–0.65	Yes	Expansion starts 500 million years ago or earlier
E. Lubimova	1967	0.03–0.05	No	Denies varying G
S. W. Carey	~1970	5–8	No	Accelerating expansion
F. Hoyle and J. Narlikar	1972	~0.1	Yes	Hoyle–Narlikar G(t) theory
J. Dooley	1973	0.5–0.6	No	Not convinced of expansion
P. Wesson	1973	~16	(No)	Expansion slowing down; possibly matter creation
H. Owen	1976	~0.45	No	Exponential growth since 600 million years ago
V. Canuto et al.	1977	0.2–0.3	(Yes)	Assumes matter creation
M. McElhinny	1978	<0.13	No	Denies expansion and varying G
S. Yabushita	1982	0.07 or 0.25	Yes	Depends on matter creation
X. Wu et al.	2011	0.1 ± 0.2	No	No present expansion

The unit of the average increase in the Earth's radius given in column 3 is mm year^{-1}. Column 4 refers to the role played by the G(t) hypothesis in either the Dirac–Jordan version or some other version. If the assumed expansion rested on or made use of the hypothesis it is marked with "yes," while "(yes)" means that the author just considered the hypothesis to be a possible contributing cause

they denied that these processes implied continental drift. Paul Lowman, a geologist at NASA's Goddard Space Flight Center, was in favour of combining plate tectonics with fixed rather than drifting continents. Aware of the expansion alternative and the varying-G mechanism, he found the expanding Earth theory to be "stimulating but quite unlikely."[175]

After the demise of the expanding Earth, or rather the marginalization of it, the decreasing-G hypothesis lived on in a variety of astronomical and physical contexts. It still does. After all, theories based on the G(t) conjecture never relied

---

[175] Lowman (1983).

crucially on Earth expansion. To mention but one example from the early period, in 1976 Dirac's conjecture turned up as a possible solution to what is presently known as the dark matter problem but at the time was usually called the problem of missing matter. This problem—essentially that the amount of visible matter was insufficient to keep galactic clusters gravitationally together—was first pointed out by Zwicky in 1933, and in the 1970s it attracted much attention in astronomy and cosmology. According to one astronomer, Dirac's "hypothesis of a variable gravitational constant can completely solve the 'missing mass' problem, and promises to assist the study of the formation, structure and evolution of galaxies."[176] Although this turned out to be an exaggerated claim and a blind alley, the consequences of the G (t) hypothesis were explored in many other ways.

The de facto refutation of Dirac's hypothesis in the mid-1980s did not have any major effect on the popularity of varying-G cosmologies among physicists. For example, in an attempt to explain various problems in the inflation theory of the very early universe, in 1989 Paul Steinhardt and his collaborator Daile La proposed an "extended inflationary model" based on a slight modification of the Jordan–Brans–Dicke theory of gravitation. While this model included varying gravity, it did not rely on the LNH hypothesis.[177] A changing gravitational constant has also been suggested as a possible solution of the problem of dark energy that was highlighted with the discovery of the accelerating universe in the late 1990s. Some versions of superstring theory and related many-dimensional unified theories predict that G and other constants of nature change in time. In short, although Dirac's version of the $G(t)$ hypothesis is dead, the general idea of G varying in time is not.

Dirac tended to believe that the Large Numbers Hypothesis and its associated G (t) hypothesis were too beautiful to be wrong, but perhaps it is better to say that they are too beautiful to be scrapped. At any rate, the general claim that some of the fundamental constants of nature may vary in time continues to be a subject of interest among physicists and cosmologists.[178]

---

[176] Lewis (1976).

[177] La and Steinhardt (1989). Paul Steinhardt, one of the founders of the inflationary scenario, later concluded that inflation was wrong. In 2001 he proposed as an alternative to the inflation theory a cyclic model of the universe which he since then has developed in various versions and still defends. See Kragh (2011), pp. 202–208.

[178] See Uzan (2003). During the last two decades the possibility of a varying speed of light, a hypothesis that neither Dirac nor Jordan considered, has attracted much attention.

Bibliography

L.T. Aldrich et al., (eds.), *Cosmological and Geological Implications of Isotope Ratio Variations.* Nuclear Science Series, Report 23. (National Research Council, Washington, DC, 1958)

E. Amaldi, *The Adventurous Life of Friedrich Georg Houtermans, Physicist (1903–1996)* (Springer, Berlin, 2012)

E. Appleton, Geomagnetism and the ionosphere. *Scientific Papers Presented to Max Born* (Oliver and Boyd, Edinburgh, 1953), pp. 1–12

R.L. Armstrong, Control of sea level relative to the continents. Nature **221**, 1042–1043 (1969)

F.L. Arnot, Cosmological theory. Nature **141**, 1142–1143 (1938)

F.L. Arnot, *Time and the Universe* (Australasian Medical Publishing, Sydney, 1941)

Y. Balashov, Uniformitarianism in cosmology: background and philosophical implications of the steady-state theory. Stud. Hist. Phil. Sci. **25**, 933–958 (1994)

V. Bargmann, Relativity. Rev. Mod. Phys. **29**, 161–174 (1957)

C.H. Barnett, A suggested reconstruction of the land masses of the Earth as a complete crust. Nature **195**, 447–448 (1962)

C.H. Barnett, Oceanic rises in relation to the expanding Earth hypothesis. Nature **221**, 1043–1044 (1969)

J.M. Barnothy, B.M. Tinsley, A critique of Hoyle and Narlikar's new cosmology. Astrophys. J. **182**, 343–349 (1973)

J.D. Barrow, *From Alpha to Omega: The Constants of Nature* (Jonathan Cape, London, 2002)

J.D. Barrow, F.J. Tipler, *The Anthropic Cosmological Principle* (Clarendon Press, Oxford, 1986)

C. Barton, Marie Tharp, oceanographic cartographer, and her contributions to the revolution in the earth sciences, in *The Earth Inside and Out: Some Major Contributions to Geology in the Twentieth Century*, ed. by D.R. Oldroyd (The Geological Society, Bransmill Lane, Bath, 2002), pp. 215–228

A.E. Beck, An expanding Earth with loss of gravitational potential energy. Nature **185**, 677–678 (1960)

A.E. Beck, Energy changes in an expanding Earth, in *The Application of Modern Physics to the Earth and Planetary Interiors*, ed. by S.K. Runcorn (Wiley Interscience, London, 1969), pp. 77–86

G.F. Becker, Relations of radioactivity to cosmogony and geology. Bull. Geol. Soc. Am. **19**, 113–146 (1908)

W.D. Beiglböck, Pascual Jordan: Schriftenverzeichnis, in *Pascual Jordan (1902–1980). Mainzer Symposium zum 100. Geburtstag*, (Max Planck Institute for the History of Science, Berlin, 2007), pp. 47–68, Preprint no. 329. http://www.mpiwg-berlin.mpg.de/en/resources/preprints.html

© Springer International Publishing Switzerland 2016
H. Kragh, *Varying Gravity*, Science Networks. Historical Studies 54,
DOI 10.1007/978-3-319-24379-5

V.V. Beloussov, Against the hypothesis of ocean-floor spreading. Tectonophysics **9**, 489–511 (1970)

P.L. Bender et al., The lunar laser ranging experiment. Science **182**, 229–238 (1973)

P.G. Bergmann, Unified field theory with fifteen field variables. Ann. Math. **49**, 255–264 (1948)

P.G. Bergmann, Summary of the Chapel Hill conference. Rev. Mod. Phys. **29**, 352–354 (1957)

R.H. Beyler, Targeting the organism: the scientific and cultural context of Pascual Jordan's quantum biology, 1932–1947. Isis **87**, 248–273 (1996)

R.H. Beyler, Ernst Pascual Jordan: freedom vs. materialism, in *Eminent Lives in Twentieth-Century Science & Religion*, ed. by N.A. Rupke (Peter Lang, Frankfurt am Main, 2009), pp. 233–252

R.H. Beyler, *From Positivism to Organicism: Pascual Jordan's Interpretations of Modern Physics in Cultural Context.* Ph.D. thesis, Harvard University, 1994

H.-J. Binge, Vulkanismus und Intrusionen als Folge der Zeitabhängigkeit von κ in der Jordanschen Kosmologie. Zeitschrift für Naturforschung A **10**, 900 (1955)

H.-J. Binge, *Folgerungen der Diracschen Hypothese für die Physik des Erdkörpers.* Unpublished dissertation, Hamburg University, 1962

F. Birch, On the possibility of large changes in the Earth's volume. Phys. Earth Planet. Inter. **1**, 141–147 (1968)

P.M.S. Blackett, Instability of the mesotron and the gravitational constant. Nature **144**, 30 (1939)

P.M.S. Blackett, Cosmic rays: recent developments. Proc. Phys. Soc. **53**, 203–213 (1941)

J.T. Blackmore, Is 'Planck's principle' true? Br. J. Philos. Sci. **29**, 347–349 (1978)

G.M. Blake, The rate of change of G. Mon. Not. R. Astron. Soc. **178**, 41P–43P (1977)

G.M. Blake, The Large Numbers Hypothesis and the rotation of the Earth. Mon. Not. R. Astron. Soc. **185**, 399–407 (1978)

H. Bondi, *Cosmology* (Cambridge University Press, Cambridge, 1952)

H. Bondi, T. Gold, On the damping of the free nutation of the Earth. Mon. Not. R. Astron. Soc. **115**, 41–46 (1955)

C.V. Boys, On the Newtonian constant of gravitation. Nature **50**, 330–334 (1894)

C.H. Brans, Gravity and the tenacious scalar field, in *On Einstein's Path: Essays in Honor of Engelbert Schucking*, ed. by A. Harvey (Springer, New York, 1999), pp. 121–138

C.H. Brans, R.H. Dicke, Mach's principle and a relativistic theory of gravitation. Phys. Rev. **124**, 925–935 (1961)

C.H. Brans, Jordan-Brans-Dicke theory. *Scholarpedia* **9** (4), 31358 (2014). http://www.scholarpedia.org/article/Jordan-Brans-Dicke_Theory

C.H. Brans, Varying Newton's constant: a personal history of scalar-tensor theories, in *Einstein Online* **04**, 1002 (2010). http://www.einstein-online.info/spotlights/scalar-tensor/?searchterm=Brans

C.H. Brans, *Mach's Principle and a Varying Gravitational Constant.* Unpublished Ph.D. thesis, Princeton University, 1961, http://loyno.edu/~brans/theses/PHD-thesis-brans-1961.pdf

V.S. Brezhnev, D.D. Ivanenko, B.N. Frolov, A possible interpretation of Dirac's hypothesis on the decrease in the gravitational constant based on a new solution of Einstein's equations. Sov. Phys. J. **9**(6), 67–68 (1966)

D.R. Brill, Review of Jordan's extended theory of gravitation, in *Evidence for Gravitational Theories*, ed. by C. Møller (Academic Press, New York, 1962), pp. 50–68

S.G. Brush, Is the Earth too old? The impact of geochronology on cosmology, 1929–1952, in *The Age of the Earth: From 4004 BC to AD 2002*, ed. by C.L. Lewis, S.J. Knell (Geological Society, London, 2001), pp. 157–175

S.G. Brush, *A History of Modern Planetary Science: Nebulous Earth* (Cambridge University Press, Cambridge, 1996a)

S.G. Brush, *A History of Modern Planetary Science: The Age of the Earth and the Evolution of the Elements from Lyell to Patterson* (Cambridge University Press, Cambridge, 1996b)

S.G. Brush, *A History of Modern Planetary Science: Fruitful Encounters* (Cambridge University Press, Cambridge, 1996c)

P. Buckley, F. David Peat, *A Question of Physics: Conversations in Physics and Biology* (Routledge & Kegan Paul, London, 1979)

K.E. Bullen, Compressibility-pressure hypothesis and the Earth's inner core. Mon. Not. R. Astron. Soc. **5**, 355–368 (1949)

K.E. Bullen, *The Earth's Density* (Chapman and Hall, London, 1975)

V. Canuto, The Earth's radius and the *G* variation. Nature **290**, 739–744 (1981)

V. Canuto, S.-H. Hsieh, The 3 K blackbody radiation, Dirac's large numbers hypothesis, and scale-covariant cosmology. Astrophys. J. **224**, 302–307 (1978)

V. Canuto, S.-H. Hsieh, Cosmological variation of *G* and the solar luminosity. Astrophys. J. **237**, 613–615 (1980a)

V. Canuto, S.-H. Hsieh, Primordial nucleosynthesis and Dirac's Large Numbers Hypothesis. Astrophys. J. **239**, L91 (1980b)

V. Canuto, J. Lodenquai, Dirac cosmology. Astrophys. J. **211**, 342–356 (1977)

V. Canuto, P.J. Adams, E. Tsiang, Crystal structure and Dirac's large numbers hypothesis. Nature **261**, 438 (1976)

V. Canuto, P.J. Adams, S.-H. Hsieh, E. Tsiang, Scale-covariant theory of gravitation and astrophysical applications. Phys. Rev. D **16**, 1643–1663 (1977)

S.W. Carey, A tectonic approach to continental drift, in *Continental Drift: A Symposium*, ed. by S.W. Carey (University of Tasmania, Hobart, TAS, 1958), pp. 177–355

S.W. Carey, Palæomagnetic evidence relevant to a change in the Earth's radius. Nature **190**, 36 (1961)

S.W. Carey, The expanding Earth—An essay review. Earth-Sci. Rev. **11**, 105–143 (1975)

S.W. Carey, *The Expanding Earth* (Elsevier, Amsterdam, 1976)

S.W. Carey, Earth expansion and the null universe, in *The Expanding Earth, A Symposium*, ed. by S. Warren Carey (University of Tasmania, Hobart, TAS, 1983), pp. 367–374

S.W. Carey, *Theories of the Earth and Universe: A History of Dogma in the Earth Sciences* (Stanford University Press, Stanford, CA, 1988)

S.W. Carey, A philosophy of the Earth and the universe. *Papers and Proceedings of the Royal Society of Tasmania*, vol. 112 (1978), http://eprints.utas.edu.au/14186/1/1978_Carey_Philosophy.pdf

B. Carter, Large number coincidences and the anthropic principle in cosmology, in *Confrontations of Cosmological Theories with Observational Data*, ed. by M.S. Longair (Reidel, Dordrecht, 1973), pp. 291–298

B. Carter, The anthropic principle: self-selection as an adjunct to natural selection, in *Cosmic Perspectives*, ed. by S.K. Biswas, D.C.V. Mallik, C.V. Vishveshwara (Cambridge University Press, Cambridge, 1989), pp. 185–206

S. Chandrasekhar, The cosmological constants. Nature **139**, 757–758 (1937)

C.-W. Chin, R. Stothers, Solar test of Dirac's large numbers hypothesis. Nature **254**, 206–207 (1975)

T.L. Chow, The variability of the gravitational constant. Lettere al Nuovo Cimento **31**, 119–120 (1981)

A.M. Clerke, *The System of the Stars* (Longmans, Green & Co., London, 1890)

I.B. Cohen, *Revolution in Science* (Harvard University Press, Cambridge, MA, 1985)

I.B. Cohen (ed.), *Isaac Newton: The Principia* (University of California Press, Berkeley, 1999)

R.S. Cohen, Epistemology and cosmology: E. A. Milne's theory of relativity. Rev. Metaphys. **3**, 385–405 (1949–50)

M.A. Cook, A.J. Eardley, Energy requirement in terrestrial expansion. J. Geophys. Res. **66**, 3907–3912 (1961)

P. Couderc, *The Expansion of the Universe* (Faber and Faber, London, 1952)

A.V. Cox, *Plate Tectonics and Geomagnetic Reversals* (Freeman, San Francisco, 1973)

A.V. Cox, R.R. Doell, Palæomagnetic evidence to a change in the Earth's radius. Nature **189**, 45–47 (1961)

K.M. Creer, An expanding Earth? Nature **205**, 539–544 (1965a)

K.M. Creer, Tracking the Earth's continents. Discovery (Popular Journal of Knowledge) **26** (February), 34–40 (1965b)

J. Croll, *Discussions on Climate and Cosmology* (A. and C. Black, Edinburgh, 1885)

J. Croll, *Stellar Evolution and its Relation to Geological Time* (Edward Stanford, London, 1889)

D.J. Crossley, R.K. Stevens, Expansion of the Earth due to a secular change in *G*—evidence from Mercury. Can. J. Earth Sci. **13**, 1723–1725 (1976)

J. Darius, Rethinking the universe. New Scientist **53**(2 March), 482–483 (1972)

B. Davis, A suggestive relation between the gravitational constant and the constants of the ether. Science **19**, 928–929 (1904)

V. De Sabatta, P. Rizzati, A relation between the periodicity of earthquakes and the variation of gravitational constant. Lettere al Nuovo Cimento **20**, 117–120 (1977)

D.S. Dearborn, D.N. Schramm, Limits on variation of *G* from clusters of galaxies. Nature **247**, 441–443 (1974)

R. Dearnley, Orogenic fold-belts, convection and expansion of the Earth. Nature **206**, 1284–1290 (1965)

R. Dearnley, Orogenic fold-belts and a hypothesis of Earth evolution. Phys. Chem. Earth **7**, 1–114 (1966)

R. Dearnley, Crustal tectonic evidence for Earth expansion, in *The Application of Modern Physics to the Earth and Planetary Interiors*, ed. by S.K. Runcorn (Wiley Interscience, London, 1969), pp. 103–110

R. Dehm, Geologisches Erdalter und astrophysikalisches Weltalter. Naturwissenschaften **36**, 166–171 (1949)

H. Dehnen, H. Hönl, Informationen über das Universum aus antipodisch beobachteten Radioquellen. Die Naturwissenschaften **55**, 413–415 (1968)

H. Dehnen, H. Hönl, Astrophysical consequences of Dirac's hypothesis of a variable gravitational number. Astrophys. J. **155**, L35–L42 (1969)

J.G. Dennis, Fitting the continents. Nature **196**, 364 (1962)

B. DeWitt, Quantum gravity: yesterday and today. Gen. Relativ. Gravit. **41**, 413–419 (2009)

C.M. DeWitt, B. DeWitt (eds.), *Relativity, Groups and Topology* (Blackie and Son, London, 1964)

C.M. DeWitt, D. Rickles (eds.), *The Role of Gravitation in Physics: Report from the 1957 Chapel Hill Conference.* (Edition Open Access, Berlin, 2011). http://edition-open-access.de/sources/5/index.html

R.H. Dicke, Principle of equivalence and the weak interactions. Rev. Mod. Phys. **29**, 355–362 (1957a)

R.H. Dicke, Gravitation without a principle of equivalence. Rev. Mod. Phys. **29**, 363–376 (1957b)

R.H. Dicke, Gravitation—An enigma. Am. Sci. **47**, 25–40 (1959a)

R.H. Dicke, Dirac's cosmology and the dating of meteorites. Nature **183**, 170–171 (1959b)

R.H. Dicke, New research on old gravitation. Science **129**, 621–624 (1959c)

R.H. Dicke, The nature of gravitation, in *Science in Space*, ed. by L.V. Berkner, H. Odishaw (McGraw-Hill, New York, 1961a), pp. 91–120

R.H. Dicke, New thinking about gravitation. New Scientist **42**(28 December), 795–797 (1961b)

R.H. Dicke, The Earth and cosmology. Science **138**, 653–664 (1962a)

R.H. Dicke, Implications for cosmology of stellar and galactic evolution rates. Rev. Mod. Phys. **34**, 110–122 (1962b)

R.H. Dicke, The many faces of Mach, in *Gravitation and Relativity*, ed. by H.-Y. Chiu, W.F. Hoffmann (W. A. Benjamin, New York, 1964a), pp. 121–141

R.H. Dicke, The significance for the solar system of time-varying gravitation, in *Gravitation and Relativity*, ed. by H.-Y. Chiu, W.F. Hoffmann (W. A. Benjamin, New York, 1964b), pp. 142–174

R.H. Dicke, Possible effects on the solar system of φ waves if they exist, in *Gravitation and Relativity*, ed. by H.-Y. Chiu, W.F. Hoffmann (W. A. Benjamin, New York, 1964c), pp. 241–257

R.H. Dicke, The secular acceleration of the Earth's rotation and cosmology, in *The Earth-Moon System*, ed. by B.G. Marsden, A.G.W. Cameron (Plenum Press, New York, 1966), pp. 98–164

R.H. Dicke, Scalar-tensor gravitation and the cosmic fireball. Astrophys. J. **152**, 1–24 (1968)

R.H. Dicke, Average acceleration of the Earth's rotation and the viscosity of the deep mantle. J. Geophys. Res. **74**, 5895–5902 (1969)

R.H. Dicke, Dirac's cosmology and Mach's principle. Nature **192**, 440–441 (1961a). Reprinted in J. Leslie (ed.), *Physical Cosmology and Philosophy* (Macmillan, New York, 1960), pp. 121–124

R.H. Dicke, P.J.E. Peebles, Gravitation and space science. Space Sci. Rev. **4**, 419–460 (1965)

R.H. Dicke, The experimental basis of Einstein's theory, in *The Role of Gravitation in Physics: Report from the 1957 Chapel Hill Conference*, eds. by C.M. DeWitt, D. Rickles, (Edition Open Access, Berlin, 2011), pp. 51–60. http://edition-open-access.de/sources/5/index.html

R.H. Dicke, P. James, E. Peebles, P.G. Roll, D.T. Wilkinson, Cosmic black-body radiation. Astrophys. J. **142**, 414–419 (1965)

R.S. Dietz, Passive continents, spreading sea floors and continental rises: a reply. Am. J. Sci. **265**, 231–237 (1967)

H. Dingle, Modern Aristotelianism. Nature **139**, 784–786 (1937)

P.A.M. Dirac, The cosmological constants. Nature **139**, 323 (1937)

P.A.M. Dirac, A new basis for cosmology. Proc. R. Soc. A **165**, 199–208 (1938)

P.A.M. Dirac, The relation between mathematics and physics. Proc. R. Soc. (Edinburgh) **59**, 122–129 (1939)

P.A.M. Dirac, The variability of the gravitational constant, in *Cosmology, Fusion, and Other Matters: George Gamow Memorial Volume*, ed. by F. Reines (Adam Hilger, London, 1972), pp. 56–59

P.A.M. Dirac, Fundamental constants and their development in time, in *The Physicist's Conception of Nature*, ed. by J. Mehra (Reidel, Dordrecht, 1973a), pp. 45–59

P.A.M. Dirac, Long range forces and broken symmetries. Proc. R. Soc. A **333**, 403–418 (1973b)

P.A.M. Dirac, New ideas of space and time. Naturwissenschaften **60**, 529–531 (1973c)

P.A.M. Dirac, Evolutionary cosmology. Pontifica Academia Scientiarum, Commentarii **11**(46), 1–15 (1973d)

P.A.M. Dirac, Cosmological models and the Large Number hypothesis. Proc. R. Soc. A **338**, 439–446 (1974)

P.A.M. Dirac, The Large Numbers Hypothesis and its consequences, in *Theories and Experiments in High-Energy Physics*, ed. by A. Perlmutter, S.M. Widmayer (Plenum Press, New York, 1975), pp. 443–456

P.A.M. Dirac, Consequences of varying *G*, in *Current Trends in the Theory of Fields*, ed. by J.E. Lannutti, P.K. Williams (AIP Conference Proceedings, New York, 1978a), pp. 169–174

P.A.M. Dirac, Cosmology and the gravitational constant, in *Directions in Physics*, ed. by P.A.M. Dirac (Wiley, New York, 1978b), pp. 71–92

P.A.M. Dirac, The Large Numbers Hypothesis and the cosmological variation of the gravitational constant, in *On the Measurement of Cosmological Variations of the Gravitational Constant*, ed. by L. Halpern (University of Florida Press, Miami, 1978c), pp. 3–20

P.A.M. Dirac, The Large Numbers Hypothesis and the Einstein theory of gravitation. Proc. R. Soc. A **365**, 19–30 (1979)

P.A.M. Dirac, The early years of relativity, in *Albert Einstein, Historical and Cultural Perspectives*, ed. by G. Holton, Y. Elkana (Princeton University Press, Princeton, 1982), pp. 79–90

R.E. Doel, *Solar System Astronomy in America: Communities, Patronage, and Interdisciplinary Research, 1920–1960* (Cambridge University Press, Cambridge, 1996)

J.C. Dooley, Is the Earth expanding? Search **4**(1–2), 9–15 (1973)

S. Ducheyne, Testing universal gravitation in the laboratory, or the significance of research of the mean density of the Earth and big *G*, 1798–1898: changing pursuits and long-term methodological-experimental continuity. Arch. Hist. Exact Sci. **65**, 181–227 (2011)

F. Dyson, Variation of constants, in *Current Trends in the Theory of Fields*, ed. by J.E. Lannutti, P.K. Williams (AIP Conference Proceedings, New York, 1978), pp. 163–168

F. Dyson, The fundamental constants and their time variation, in *Aspects of Quantum Theory*, eds. by A. Salam, E. P. Wigner, (1972), pp. 213–236

A.C. Economos, The largest land mammal. J. Theor. Biol. **89**, 211–215 (1981)

A.S. Eddington, The borderland of astronomy and geology. Nature **111**, 18–21 (1923)

A.S. Eddington, The cosmological controversy. Sci. Prog. **34**, 225–236 (1939)

L. Egyed, Determination of changes in the dimension of the Earth from palæogeographical data. Nature **173**, 534 (1956a)

L. Egyed, The change of the Earth's dimensions determined from paleogeographical data. Geofisica Pura e Applicata **33**, 42–48 (1956b)

L. Egyed, A new dynamic conception of the internal constitution of the Earth. Geol. Rundsch. **46**, 101–121 (1957)

L. Egyed, On the origin and constitution of the upper part of the Earth's mantle. Geol. Rundsch. **50**, 251–258 (1960a)

L. Egyed, Some remarks on continental drift. Geofisica Pura e Applicata **45**, 115–116 (1960b)

L. Egyed, Dirac's cosmology and the origin of the solar system. Nature **186**, 621–622 (1960c)

L. Egyed, The effect of internal processes and paleoclimates. Ann. N. Y. Acad. Sci. **95**, 508–512 (1961a)

L. Egyed, The expanding Earth. Trans. N. Y. Acad. Sci. **23**, 424–432 (1961b)

L. Egyed, The expanding Earth? Nature **197**, 1059–1060 (1963)

L. Egyed, Vom Aufbau der Erde, in *Die Erde*, ed. A. Tasnádi-Kubacska (Urania-Verlag, Leipzig, 1965), pp. 48–103

L. Egyed, The slow expansion hypothesis, in *The Application of Modern Physics to the Earth and Planetary Interiors*, ed. by S.K. Runcorn (Wiley Interscience, London, 1969a), pp. 65–75

L. Egyed, *Physik der Festen Erde* (Teubner, Leipzig, 1969b)

L. Egyed, L. Stegena, Physical background of a dynamical Earth model. Z. Geophys. **24**, 260–267 (1958)

J. Ehlers, E. Schücking, 'Aber Jordan war der erste': Zur Erinnerung an Pascual Jordan (1902–1980). Phys. J. **1**(11), 71–74 (2002)

W. Eichendorf, M. Reinhardt, How constant are fundamental physical quantities? Zeitschrift für Naturforschung A **32a**, 532–537 (1977)

G. Ellis, Editorial note. Gen. Relativ. Gravit. **41**, 2179–2189 (2009)

W.M. Elsasser, Sea-flooor spreading as thermal convection. J. Geophys. Res. **76**, 1101–1112 (1971)

H.T. Engelhardt, A.L. Caplan (eds.), *Scientific Controversies: Case Studies in the Resolution and Closure of Disputes in Science and Technology* (Cambridge University Press, Cambridge, 1987)

D. Ezer, A.G.W. Cameron, Solar evolution with varying *G*. Can. J. Phys. **44**, 593–615 (1966)

R.W. Fairbridge, Thoughts about an expanding globe, in *Advancing Frontiers in Geology and Geophysics*, ed. by A.P. Subramanian, S. Balakrishna (Indian Geophysical Union, Hyderabad, 1964), pp. 59–88

R.W. Fairbridge, Endospheres and interzonal coupling. Ann. N. Y. Acad. Sci. **140**, 133–148 (1966)

D. Falik, Primordial nucleosynthesis and Dirac's Large Numbers Hypothesis. Astrophys. J. **231**, L1 (1979)

G. Feulner, The faint young Sun problem. Rev. Geophys. **50**, RG2006 (2012)

M. Fierz, Über die physikalische Deutung der erweiterten Gravitationstheorie P. Jordans. Helvetica Physica Acta **29**, 128–134 (1956)

J.R. Fleming, T. C. Chamberlin, climate change, and cosmogony. Stud. Hist. Philos. Mod. Phys. **31**, 293–308 (2000)

J.R. Fleming, James Croll in context: the encounter between climate dynamics and geology in the second half of the nineteenth century, in *Milutin Milankovitch Anniversary Symposium:*

Paleoclimate and the Earth Climate System, ed. by A. Berger, M. Ercegovac, F. Mesinger (Serbian Academy of Sciences and Arts, Belgrade, 2005), pp. 13–20

H.R. Frankel, Alfred Wegener and the specialists. Centaurus **20**, 305–324 (1976)

H.R. Frankel, *The Continental Drift Controversy: Wegener and the Early Debate* (Cambridge University Press, Cambridge, 2012a)

H.R. Frankel, *The Continental Drift Controversy: Paleomagnetism and Confirmation of Drift* (Cambridge University Press, Cambridge, 2012b)

H.R. Frankel, *The Continental Drift Controversy: Introduction of Seafloor Spreading* (Cambridge University Press, Cambridge, 2012c)

H.R. Frankel, *The Continental Drift Controversy: Evolution into Plate Tectonics* (Cambridge University Press, Cambridge, 2012d)

G. Gamow, Any physics tomorrow? Phys. Today **2**(1), 16–21 (1949)

G. Gamow, *Gravity: Classic and Modern Views* (Heinemann, London, 1962)

G. Gamow, Electricity, gravity and cosmology. Phys. Rev. Lett. **19**, 759–761 (1967a)

G. Gamow, History of the universe. Science **158**, 766–769 (1967b)

G. Gamow, Does gravity change with time? Proc. Natl. Acad. Sci. U.S.A. **57**, 187–193 (1967c)

G. Gamow, E. Teller, On the origin of great nebulae. Phys. Rev. **55**, 654–657 (1939)

H. Gerstenkorn, Veränderungen der Erde-Monde-System durch Gezeitenreibung in der Vergangenheit bei zeitabhängiger Gravitationskonstante. Z. Astrophys. **42**, 137–155 (1957)

C. Gilbert, Dirac's cosmology and the general theory of relativity. Mon. Not. R. Astron. Soc. **116**, 684–690 (1956)

C. Gilbert, The general theory of relativity and Newton's law of gravitation. Nature **179**, 270 (1957)

C. Gilbert, Dirac's cosmology. Nature **192**, 57 (1961)

G.T. Gillies, The Newtonian gravitational constant: recent measurements and related studies. Rep. Prog. Phys. **60**, 151–225 (1997)

G.T. Gillies, *The Newtonian Gravitational Constant: An Index of Measurements*. Report BIPM-83/1. Sèvres, (Bureau International des Poids et Mesures, France, 1983). http://www.bipm.org/utils/common/pdf/rapportBIPM/1983/01.pdf

J.H. Gittus, Dirac's large numbers hypothesis and the structure of rocks. Proc. R Soc. A **343**, 155–158 (1975)

H. Glashoff, *Endogene Dynamik der Erde und die Diracsche Hypothese* (Mathematisch-Naturwissenschaftlichen Klasse, Akademie der Wissenschaften und der Literatur in Mainz, 1966), p. 34

I.S. Glass, Jacob Karl Ernst Halm (1865–1944). Mon. Notes Astron. Soc. South Africa **73**, 14–23 (2014)

H. Goenner, Some remarks on the genesis of scalar-tensor theories. Gen. Relativ. Gravit. **44**, 2077–2097 (2012)

T. Gold, Instability of the Earth's axis of rotation. Nature **175**, 526–529 (1955)

J.N. Goldberg, US Air Force support of general relativity, 1956–1972, in *Studies in the History of General Relativity*, ed. by J. Eisenstaedt, A.J. Kox (Birkhäuser, Boston, 1992), pp. 89–102

J. Greenberg, *The Problem of the Earth's Shape from Newton to Clairaut* (Cambridge University Press, Cambridge, 1995)

J. Gregory, *Fred Hoyle's Universe* (Oxford University Press, Oxford, 2005)

J.S. Grimes, *Outlines of Geonomy: A Treatise on the Physical Laws of the Earth and the Creation of the Continents* (Phillips, Sampson & Company, Boston, 1858)

A.E. Haas, An attempt to a purely theoretical derivation of the mass of the universe. Phys. Rev. **49**, 411–412 (1936)

J.B.S. Haldane, A quantum theory of the origin of the solar system. Nature **155**, 133–135 (1945a)

J.B.S. Haldane, A new theory of the past. Am. Sci. **33**, 129–145 (1945b)

A. Hallam, Re-evaluation of the palaeogeographic argument for an expanding Earth. Nature **232**, 180–182 (1971)

A. Hallam, *A Revolution in the Earth Sciences: From Continental Drift to Plate Tectonics* (Clarendon Press, Oxford, 1973)

A. Hallam, The unlikelihood of an expanding Earth. Geol. Mag. **121**, 653–655 (1984)

J.K.E. Halm, An astronomical aspect of the evolution of the Earth. J. Astron. Soc. South Africa **4**, 1–28 (1935a)

J.K.E. Halm, On the theory of an 'expanding universe'. J. Astron. Soc. South Africa **4**, 29–31 (1935b)

W.B. Harland, An expanding Earth? Nature **278**, 12–13 (1979)

E.R. Harrison, Cosmic numbers. Nature **197**, 1257–1259 (1963)

O.H.L. Heckmann, E. Schücking, Andere kosmologische Theorien, in *Handbuch der Physik*, ed. by S. Flügge, vol. 53 (Springer, Berlin, 1959), pp. 520–357

B.C. Heezen, The rift in the ocean floor. Sci. Am. **203**(October), 98–110 (1960)

B.C. Heezen, The deep-sea floor, in *Continental Drift*, ed. by S.K. Runcorn (Academic Press, New York, 1962), pp. 235–288

R.W. Hellings et al., Experimental test of the variability of *G* using Viking Lander Ranging Data. Phys. Rev. Lett. **51**, 1609–1612 (1983)

V. Herzen, P. Richard, Surface heat flow and some implications for the mantle, in *The Earth's Mantle*, ed. by T.F. Gaskell (Academic Press, London, 1967), pp. 197–231

O.C. Hilgenberg, Paläopollagen der Erde. N. Jb. Geol. Paläont. **116**, 1–56 (1962)

A. Holmes, *The Age of the Earth* (Harper & Brothers, London, 1913)

A. Holmes, Radioactivity and Earth history. Geogr. J. **65**, 528–532 (1925)

A. Holmes, *Principles of Physical Geology* (Ronald Press, New York, 1965)

H. Hönl, Zwei Bemerkungen zur kosmologischen Problem. Ann. Phys. **6**, 169–176 (1949)

H. Hönl, H. Dehnen, Erlaubt die 3° Kelvin-Strahlung Rückschlüsse auf eine konstante oder veränderliche Gravitationszahl? Z. Astrophys. **68**, 181–189 (1968)

J. Hospers, S. Van Andel, Palaeomagnetism and the hypothesis of an expanding Earth. Tectonophysics **5**, 5–24 (1967)

J. Hospers, S. Van Andel, Statistical analysis of ancient Earth radii computed from palaeomagnetic data, in *Palaeogeophysics*, ed. by S. Keith Runcorn (Academic Press, London, 1970), pp. 407–412

F.G. Houtermans, P. Jordan, Über die Annahme der zeitlichen Veränderlichkeit des β-Zerfalls und die Möglichkeiten ihrer experimentellen Prüfung. Z. Naturforsch. **1**, 125–130 (1946)

F. Hoyle, Remarks on the computation of evolutionary tracks, in *Stellar Populations*, ed. by J.K. O'Connell (Specola Vaticana, Vatican City, 1958), pp. 223–230

F. Hoyle, The history of the Earth. Q. J. R. Astron. Soc. **13**, 328–345 (1972)

F. Hoyle, *Home is Where the Wind Blows: Chapters from a Cosmologist's Life* (University Science Books, Mill Valley, CA, 1994)

F. Hoyle, R.A. Lyttleton, The effect of interstellar matter on climatic variation. Proc. Camb. Philos. Soc. **35**, 405–415 (1939)

F. Hoyle, J.V. Narlikar, A new theory of gravitation. Proc. R. Soc. A **282**, 191–207 (1964)

F. Hoyle, J.V. Narlikar, On the nature of mass. Nature **233**, 41–44 (1971)

F. Hoyle, J.V. Narlikar, Cosmological models in a conformally invariant gravitational theory, II. Mon. Not. R. Astron. Soc. **155**, 323–335 (1972)

S.W. Hurrell, *Dinosaurs and the Expanding Earth* (2011), OneOffPublishing.com (E-book)

E.A. Irving, *Paleomagnetism and Its Application to Geological and Geophysical Problems* (Wiley, New York, 1964)

W. Israel, Imploding stars, shifting continents, and the inconstancy of matter. Found. Phys. **26**, 595–616 (1996)

H. Jeffreys, *The Earth: Its Origin, History and Physical Constitution* (Cambridge University Press, Cambridge, 1924)

H.C. Joksch, Statitische Analyse der hypsometrischen Kurve der Erde. Z. Geophys. **21**, 109–112 (1955)

J. Joly, *Radioactivity and Geology: An Account of Radioactive Energy in Terrestrial History* (Constable, London, 1909)

P. Jordan, Über den positivischen Begriff der Wirklichkeit. Die Naturwissenschaften **22**, 485–490 (1934)

P. Jordan, *Die Physik des 20 Jahrhhunderts* (Vieweg, Braunschweig, 1936)

P. Jordan, Die physikalischen Weltkonstanten. Die Naturwissenschaften **25**, 513–517 (1937)

P. Jordan, Zur empirischen Kosmologie. Die Naturwissenschaften **26**, 417–421 (1938)

P. Jordan, Bemerkungen zur Kosmologie. Ann. Phys. **32**, 64–70 (1939)

P. Jordan, Über die Entstehung der Sterne. Physikalische Zeitschrift **45**(183–190), 233–244 (1944)

P. Jordan, *Die Herkunft der Sterne* (Hirzel, Stuttgart, 1947a)

P. Jordan, *Das Bild der modernen Physik* (Stromverlag Hamburg, Bergedorf, 1947b)

P. Jordan, Fünfdimensionale Kosmologie. Astronomische Nachrichten **276**, 193–208 (1948)

P. Jordan, Formation of the stars and development of the universe. Nature **164**, 637–640 (1949)

P. Jordan, *Schwerkraft und Weltall: Grundlagen der theoretischen Kosmologie* (Vieweg & Sohn, Braunschweig, 1952)

P. Jordan, *Schwerkraft und Weltall: Grundlagen der theoretischen Kosmologie*, Second revisedth edn. (Vieweg & Sohn, Braunschweig, 1955)

P. Jordan, Zum gegenwärtigen Stand der Diracschen kosmologischen Hypothesen. Z. Phys. **157**, 112–121 (1959a)

P. Jordan, Die Bedeutung der Diracschen Hypothese für die Geophysik. *Akademie der Wissenschaften und der Literatur in Mainz, Mathematisch-Naturwissenschaftlichen Klasse*, **9**, 771–795, (1959b)

P. Jordan, Zum Problem der Erdexpansion. Die Naturwissenschaften **48**, 417–425 (1961)

P. Jordan, Geophysical consequences of Dirac's hypothesis. Rev. Mod. Phys. **34**, 596–600 (1962a)

P. Jordan, Empirical confirmation of Dirac's hypothesis of diminishing gravitation, in *Recent Developments in General Relativity* (Pergamon Press, Oxford, 1962b), pp. 283–288

P. Jordan, Remarks about Ambarzumian's conception of pre-stellar matter, in *Recent Developments in General Relativity* (Pergamon Press, Oxford, 1962c), pp. 289–292

P. Jordan, *Der Naturwissenschaftler vor der Religiösen Frage: Abbruch einer Mauer* (Gerhard Stalling Verlag, Oldenburg, 1963)

P. Jordan, Four lectures about problems of cosmology, in *Cosmological Models*, ed. by A. Giáo (Instituto Gulbenkian de Ciêcia, Lisbon, 1964), pp. 101–136

P. Jordan, *Die Expansion der Erde: Folgerungen aus der Diracschen Gravitationshypothese* (Vieweg & Sohn, Braunschweig, 1966)

P. Jordan, *Über die Wolkenhülle der Venus* (Akademie der Wissenschaften und der Literatur in Mainz, Mathematisch-Naturwissenschaftlichen Klasse, 1967), pp. 43–53

P. Jordan, Bemerkungen zu der Arbeit von H. Hönl und H. Dehnen. Z. Astrophys. **68**, 201–203 (1968)

P. Jordan, On the possibility of avoiding Ramsey's hypothesis in formulating a theory of Earth expansion, in *The Application of Modern Physics to the Earth and Planetary Interiors*, ed. by S.K. Runcorn (Wiley Interscience, London, 1969a), pp. 55–62

P. Jordan, *Albert Einstein: Sein Lebenswerk und die Zukunft der Physik*. (Stuttgart, Verlag Huber, 1969b)

P. Jordan, *The Expanding Earth: Some Consequences of Dirac's Gravitation Hypothesis* (Pergamon Press, Oxford, 1971)

P. Jordan, The expanding Earth, in *The Physicist's Conception of Nature*, ed. by J. Mehra (Reidel, Dordrecht, 1973), pp. 60–70

P. Jordan, *The theory of a variable 'constant' of gravitation*. Unpublished essay to the Gravity Research Foundation (1954). http://www.gravityresearchfoundation.org/pdf/awarded/1954/jordan.pdf

P. Jordan, *Problems of Gravitation*. Mimeographed report, Aeronautical Research Laboratory, (1961b)

P. Jordan, J. Ehlers, W. Kundt, Quantitatives zur Diracschen Schwerkraft-Hypothese. Z. Phys. **178**, 501–518 (1964)

C. Jungnickel, R. McCormmach, *Cavendish: The Experimental Life* (Bucknell University Press, Cranbury, NJ, 1999)

K. Just, Zur Kosmologie mit veränderlicher Gravitationszahl. Z. Phys. **140**, 648–655 (1955)

D. Kaiser, Is ψ just a ψ? Pedagogy, practice, and the reconstitution of general relativity, 1942–1975. Stud. Hist. Philos. Mod. Phys. **29**, 321–338 (1998)

D. Kaiser, When fields collide. Sci. Am. **350**(June), 62–69 (2007)

E.R. Kanasewich, J.C. Savage, Dirac's cosmology and radioactive dating. Can. J. Phys. **41**, 1911–1922 (1963)

R.O. Kapp, *Towards a Unified Cosmology* (Hutchinson & Co., London, 1960)

V.E. Khain, Mobilism and plate tectonics in the USSR. Tectonophysics **199**, 137–148 (1991)

V.E. Khain, A.G. Ryabukhin, Russian geology and the plate tectonics revolution, in *The Earth Inside and Out: Some Major Contributions to Geology in the Twentieth Century*, ed. by D.R. Oldroyd (The Geological Society of London, London, 2002), pp. 185–198

L.C. King, *Wandering Continents and Spreading Sea Floors on an Expanding Earth* (Wiley, New York, 1983)

R.T. King et al. (eds.), *Bibliography of North American Geology, 1950–1959* (United States Government Printing Office, Washington, DC, 1965)

G.J. Kirby, The amateur scientist and the rotation of the Earth. J. Naval Sci. **1**, 242–247 (1971)

R. Klee, The revenge of Pythagoras: how a mathematical sharp practice undermines the contemporary design argument in astrophysical cosmology. Br. J. Philos. Sci. **53**, 331–354 (2002)

H.B. Klepp, Terrestrial, interplanetary and universal expansion. Nature **201**, 693 (1964)

D.S. Kothari, Cosmological and atomic constants. Nature **142**, 354–355 (1938)

H. Kragh, *Dirac: A Scientific Biography* (Cambridge University Press, Cambridge, 1990)

H. Kragh, Cosmonumerology and empiricism: the Dirac-Gamow dialogue. Astron. Q. **8**, 109–126 (1991)

H. Kragh, *Cosmology and Controversy: The Historical Development of Two Theories of the Universe* (Princeton University Press, Princeton, 1996)

H. Kragh, The electrical universe: grand cosmological theory versus mundane experiments. Perspect. Sci. **5**, 199–231 (1997)

H. Kragh, The chemistry of the universe: Historical roots of modern cosmochemistry. Ann. Sci. **57**, 353–368 (2000)

H. Kragh, From geochemistry to cosmochemistry: the origin of a scientific discipline, in *Chemical Sciences in the 20th Century*, ed. by C. Reinhardt (Wiley-VCH, Weinheim, 2001), pp. 160–190

H. Kragh, *Matter and Spirit in the Universe: Scientific and Religious Preludes to Modern Cosmology* (Imperial College Press, London, 2004)

H. Kragh, Cosmic radioactivity and the age of the universe, 1900–1930. J. Hist. Astron. **38**, 393–412 (2007a)

H. Kragh, *Conceptions of Cosmos. From Myths to the Accelerating Universe: A History of Cosmology* (Oxford University Press, Oxford, 2007b)

H. Kragh, *Entropic Creation: Religious Contexts of Thermodynamics and Cosmology* (Ashgate, Aldershot, 2008)

H. Kragh, *Higher Speculations: Grand Theories and Failed Revolutions in Physics and Cosmology* (Oxford University Press, Oxford, 2011)

H. Kragh, Zöllner's universe. Phys. Perspect. **14**, 392–420 (2012)

H. Kragh, Nordic cosmogonies: Birkeland, Arrhenius and fin-de-siècle cosmical physics. Eur. Phys. J. H **38**, 549–572 (2013)

H. Kragh, Naming the big bang. Hist. Stud Nat. Sci. **44**, 3–36 (2014a)

H. Kragh, The science of the universe: cosmology and science education, in *International Handbook of Research in History, Philosophy and Science Teaching*, ed. by M.R. Matthews, vol. 1 (Springer, Dordrecht, 2014b), pp. 643–668

H. Kragh, Pascual Jordan, varying gravity, and the expanding Earth. Phys. Perspect. **17**, 107–134 (2015a)

H. Kragh, Expanding Earth and declining gravity: a chapter in the recent history of geophysics. Hist. Geo- Space Sci. **6**, 45–55 (2015b)

H. Kragh, *Gravitation and the Earth sciences: the contributions of Robert Dicke*. (2015c), Arxiv:1501.04293 [physics. hist-ph]

H. Kragh, *Expanding Earth and static universe: two papers of 1935*, (2015d), Arxiv:1507.08040 [physics. hist-ph]

Y. Kramarovskii, V. Chechev, Does the charge of the electron vary with the age of the universe? Soviet Physics Uspekhi **13**, 628–631 (1971)

B. Kuchowicz, Diminishing gravitation—a hitherto underrated factor in the evolution of organic life. Experientia **27**, 616 (1971)

W. Kundt, Jordan's 'excursion' into geophysics, in *Pascual Jordan (1902–1980). Mainzer Symposium zum 100. Geburtstag*, (Max Planck Institute for the History of Science, Berlin, 2007), pp. 123–132, Preprint no. 2007. http://www.mpiwg-berlin.mpg.de/en/resources/preprints.html

D. La, P.J. Steinhardt, Extended inflationary cosmology. Phys. Rev. Lett. **62**, 276–378 (1989)

R. Laudan, The recent revolution in geology and Kuhn's theory of scientific change, in *Paradigms and Revolutions: Appraisals and Applications of Thomas Kuhn's Philosophy of Science*, ed. by G. Gutting (University of Notre Dame Press, Notre Dame, 1980), pp. 284–297

R. Laudan, Redefinitions of a discipline: histories of geology and geological history, in *Functions and Uses of Disciplinary Histories*, ed. by L. Graham, W. Lepenies, P. Weingart (Reidel, Dordrecht, 1983), pp. 79–104

R. Laudan, *From Mineralogy to Geology: The Foundation of a Science, 1650–1830* (University of Chicago Press, Chicago, 1987)

H.E. Le Grand, *Drifting Continents and Shifting Theories* (Cambridge University Press, Cambridge, 1988)

G.E. Lemaître, The cosmological constant, in *Albert Einstein: Philosopher-Scientist*, ed. by P.A. Schilpp (Library of Living Philosophers, New York, 1949), pp. 437–456

L.S. Levitt, The gravitational constant at time zero. Lettere al Nuovo Cimento **29**, 23–24 (1980)

B.M. Lewis, Variable *G*: a solution to the missing mass problem. Nature **261**, 302–304 (1976)

A. Lightman, R. Brawer, *Origins: The Lives and Worlds of Modern Cosmologists* (Harvard University Press, Cambridge, MA, 1990)

P.D. Lowman, Faulting continental drift. The Sciences **23**, 34–39 (1983)

E.A. Lubimova, Theory of thermal state of the Earth's mantle, in *The Earth's Mantle*, ed. by T.F. Gaskell (Academic Press, London, 1967), pp. 231–326

C. Lyell, *Principles of Geology*, vol. 1 (John Murray, London, 1830)

R.A. Lyttleton, The structures of the terrestrial planets. Adv. Astron. Astrophys. **7**, 83–147 (1970)

R.A. Lyttleton, Relation of a contracting Earth to the apparent accelerations of the Sun and Moon. The Moon **16**, 41–58 (1976)

R.A. Lyttleton, *The Earth and its Mountains* (Wiley, New York, 1982)

R.A. Lyttleton, H. Bondi, How plate tectonics may appear to a physicist. J. Br. Astron. Assoc. **102**, 194–195 (1992)

R.A. Lyttleton, J.P. Fitch, Cosmological change of *G* and the structure of the Earth. Mon. Not. R. Astron. Soc. **180**, 471–477 (1977)

J. MacDougall et al., A comparison of terrestrial and universal expansion. Nature **199**, 1080 (1963)

F. Machado, Geological evidence for a pulsating gravitation. Nature **214**, 1317–1318 (1967)

P. Machamer, M. Pera, A. Baltas (eds.), *Scientific Controversies: Philosophical and Historical Perspectives* (Oxford University Press, New York, 2000)

A.S. Mackenzie, *The Laws of Gravitation* (American Book Company, New York, 1900)

W.D. MacMillan, Thomas Chrowder Chamberlin. Astrophys. J. **69**, 1–7 (1929)

A. Maeder, Four basic solar and stellar tests of cosmologies with variable past *G* and macroscopic masses. Astron. Astrophys. **56**, 359–367 (1977)

V.N. Mansfield, Dirac cosmologies and the microwave background. Astrophys. J. **210**, L137–L138 (1976)

J. Martinez-Frias, D. Hochberg, F. Rull, A review of the contributions of Albert Einstein to earth sciences. Naturwissenschaften **93**, 66–71 (2006)

U.B. Marvin, *Continental Drift: The Evolution of a Concept* (Smithsonian Institution Press, Washington, DC, 1973)

W. Marx, L. Bornmann, The emergence of plate tectonics and the Kuhnian model of paradigm shift. Scientometrics **94**, 595–614 (2013)

W.H. McCrea, Continual creation. Mon. Not. R. Astron. Soc. **128**, 335–343 (1964)

M.W. McElhinny, Limits to Earth expansion. Explor. Geophys. **9**, 149–152 (1978)

M.W. McElhinny, S.R. Taylor, D.J. Stevenson, Limits to the expansion of Earth, Moon, Mars and Mercury and to changes in the gravitational constant. Nature **271**, 316–321 (1978)

D.P. McKenzie, Plate tectonics and its relationship to the evolution of ideas in the geological sciences. Daedalus **106**, 97–124 (1977)

H.W. Menard, *The Ocean of Truth: A Personal History of Global Tectonics* (Princeton University Press, Princeton, 1986)

A. Mercier, M. Kervaire (eds.), *Fünfzig Jahre Relativitätstheorie* (Birkhäuser, Basel, 1956)

J. Merleau-Ponty, *La Science de l'Univers à l'Âge du Positivism* (Vrin, Paris, 1983)

R.H. Meservey, Topological inconsistency of continental drift on the present-sized Earth. Science **166**, 609–611 (1969)

A. Meskó, In memoriam László Egyed. Palaeogeogr. Palaeoclimatol. Palaeoecol. **9**, 73–75 (1971)

E.A. Milne, *Relativity, Gravitation and World-Structure* (Clarendon Press, Oxford, 1935)

W.J. Morgan, J.O. Stoner, R.E. Dicke, Periodicity of earthquakes and the invariance of the gravitational constant. J. Geophys. Res. **66**, 3831–3843 (1961)

P.M. Muller, Determination of the cosmological rate of change of G and tidal accelerations of Earth and Moon from ancient and modern astronomical data, in *On the Measurement of Cosmological Variations of the Gravitational Constant*, ed. by L. Halpern (University of Florida Press, Miami, 1978), pp. 91–116

J. Müller, L. Biskukep, Variations of the gravitational constant from lunar laser ranging data. Classical Quantum Gravity **24**, 4533–4538 (2007)

C.T. Murphy, R.H. Dicke, The effects of a decreasing gravitational constant in the interior of the Earth. Proc. Am. Philos. Soc. **108**, 224–246 (1964)

J.V. Narlikar, A.K. Kembhavi, Non-standard cosmologies, in *Handbook of Astronomy, Astrophysics and Geophysics*, eds. by V.M. Canuto, B.G. Elmegreen, vol. II: *Galaxies and Cosmology* (Gordon and Breach, New York, 1988), pp. 301–498

M.J. Newman, R.T. Rood, Implications of solar evolution for the Earth's early atmosphere. Science **198**, 1035–1037 (1977)

J. North, *The Measure of the Universe: A History of Modern Cosmology* (Oxford University Press, Oxford, 1965)

R. Nunan, The theory of an expanding Earth and the acceptability of guiding assumptions, in *Scrutinizing Science: Empirical Studies of Scientific Change*, ed. by A. Donovan, L. Laudan, R. Laudan (Kluwer Academic, Dordrecht, 1988), pp. 289–313

R. Nunan, Expanding Earth theories, in *Sciences of the Earth: An Encyclopedia of Events, People, and Phenomena*, ed. by G.A. Good, vol. 2 (Garland Publishing, New York, 1998), pp. 243–250

H. Nussbaumer, L. Bieri, *Discovering the Expanding Universe* (Cambridge University Press, Cambridge, 2009)

G.G. Nyambuya, On the expanding Earth and shrinking Moon. Int. J. Astron. Astrophys. **4**, 227–243 (2014)

M.J. Nye, *Blackett: Physics, War, and Politics in the Twentieth Century* (Harvard University Press, Cambridge, MA, 2004)

J. O'Hanlon, K.-K. Tam, Stellar ages and an extended theory of gravitation. Prog. Theor. Phys. **43**, 684–688 (1970)

D. Oldroyd, *Thinking about the Earth: A History of Ideas in Geology* (Athlone, London, 1996)

G.C. Omer, A nonhomogeneous cosmological model. Astrophys. J. **109**, 164–176 (1949)

E.J. Öpik, *The Oscillating Universe* (Mentor Book, New York, 1956)

E.J. Öpik, Solar variability and palaeoclimatic changes. Ir. Astron. J. **5**, 97–109 (1958)

E.J. Öpik, Climatic change in cosmic perspective. Icarus **4**, 289–307 (1965)

N. Oreskes, *The Rejection of Continental Drift: Theory and Method in American Earth Science* (Oxford University Press, New York, 1999)

N. Oreskes (ed.), *Plate Tectonis: An Insider's History of the Modern Theory of the Earth* (Westview Press, Cambridge, MA, 2001)

H.G. Owen, Continental displacement and expansion of the Earth during the Mesozoic and Cenozoic. Philos. Trans. R. Soc. A **281**, 223–291 (1976)

H. G. Owen, The Earth is expanding and we don't know why. *New Scientist* **65** (22 November): 27–29 (1984)

H.G. Owen, Earth expansion: some mistakes, what happened the Palaeozoic and the way ahead, in *The Earth Expansion Evidence: A Challenge for Geology, Geophysics and Astronomy*, ed. by G. Scalera, E. Boschi, S. Cwojdzinski (Istituto Nazionale di Geofisica e Vulcanologia, Rome, 2012), pp. 77–89

G. Pannella, Paleontological evidence on the Earth's rotational history since early Precambrian. Astrophys. Space Sci. **16**, 212–237 (1972)

W. Pauli, Raum, Zeit und Kausalität in der modernen Physik. *Scientia* **59**, 65–76. English translation in W. Pauli, *Writings on Physics and Philosophy*, eds. by C. P. Enz, K. von Meyenn (Springer, Berlin, 1936), pp. 95–106

W. Pauli, *Wissenschaftlicher Briefwechsel mit Bohr, Einstein, Heisenberg u.a.*, vol. 4, part 1, ed. by K. von Meyenn (Springer, Berlin, 1996)

P.J.E. Peebles, The Eötvös experiment, spatial isotropy, and generally covariant field theories of gravity. Ann. Phys. **20**, 240–260 (1962)

P.J.E. Peebles, Dicke, Robert Henry, in *New Dictionary of Scientific Biography*, ed. by N. Koertge (Tomson-Gale, Detroit, 2008), pp. 280–284

P.J.E. Peebles, R.H. Dicke, The temperature of meteorites and Dirac's cosmology and Mach's principle. J. Geophys. Res. **67**, 4063–4070 (1962a)

P.J.E. Peebles, R.H. Dicke, Cosmology and the radioactive decay ages of terrestrial rocks and meteorites. Phys. Rev. **128**, 2006–2011 (1962b)

P.J.E. Peebles, R.H. Dicke, Significance of spatial isotropy. Phys. Rev. **127**, 629–631 (1962c)

P.J.E. Peebles, D.T. Wilkinson, The primeval fireball. Sci. Am. **216**(June), 28–37 (1967)

P.J.E. Peebles, L.A. Page, R.B. Partridge, *Finding the Big Bang* (Cambridge University Press, Cambridge, 2009)

P. Pochoda, M. Schwarzschild, Variation of the gravitational constant and the evolution of the sun. Astrophys. J. **139**, 587–593 (1964)

J.H. Poynting, *Collected Scientific Papers by John Henry Poynting* (Cambridge University Press, Cambridge, 1920)

W.H. Ramsey, On the nature of the Earth's core. Geophys. J. Int. **5**(suppl. 9), 409–426 (1949)

G. Ranalli, The expansion-undation hypothesis for geotectonic evolution. Tectonophysics **11**, 261–285 (1971)

K. Rankama, T. Sahama, *Geochemistry* (University of Chicago Press, Chicago, 1950)

M. Reinhardt, Mach's principle—A critical review. *Zeitschrift für Naturforschung A* **28a**, 529–539 (1973)

P.H. Reitan, The Earth's volume change and its significance for orogenesis. J. Geol. **68**, 678–680 (1960)

A.E. Ringwood, Changes in solar luminosity and some possible terrestrial consequences. Geochim. Cosmochim. Acta **21**, 295–296 (1961)

I.W. Roxburgh, Dirac's continuous creation cosmology and the temperature of the Earth. Nature **261**, 301–302 (1976)

J.P. Rozelot, et al. A brief history of the solar oblateness, (2010) https://hal.archives-ouvertes.fr/hal-00519433/

S.K. Runcorn, Changes in the Earth's moment of inertia. Nature **204**, 823–825 (1964)

S.K. Runcorn, Mechanism of plate tectonics: mantle convection currents, plumes, gravity sliding or expansion? Tectonophysics **63**, 297–307 (1980)

S.K. Runcorn, Corals and the history of the Earth's rotation. *Sea Frontiers* **13** (January), 4–12. Reprinted in P. Cloud, (ed.), *Adventures in Earth History* (W. H. Freeman and Company, San Fransisco, 1967), pp. 190–195

C. Sagan, G. Mullen, Earth and Mars: evolution of atmospheres and surface temperatures. Science **177**, 52–56 (1972)

S. Sambursky, Static universe and nebular red shift. Phys. Rev. **52**, 335–338 (1937)

A.R. Sandage, Current problems in the extragalactic distance scale. Astrophys. J. **127**, 513–526 (1958)

G. Scalera, T. Braun, Ott Christoph Hilgenberg in twentieth-century geophysics, in *Why Expanding Earth? A Book in Honour of Ott Christoph Hilgenberg*, ed. by G. Scalera, K.-H. Jacob (Istituto Nazionale di Geofisica e Vulcanologia, Rome, 2003), pp. 25–41

G. Scalera, K.-H. Jacob (eds.), *Why Expanding Earth? A Book in Honour of Ott Christoph Hilgenberg* (Istituto Nazionale di Geofisica e Vulcanologia, Rome, 2003)

G. Scalera, E. Boschi, S. Cwojdzinski (eds.), *The Earth Expansion Evidence: A Challenge for Geology, Geophysics and Astronomy* (Istituto Nazionale di Geofisica e Vulcanologia, Rome, 2012)

E. Schatzmann, *The Origin and Evolution of the Universe* (Hutchison & Company, London, 1966). Translation of *Origine et Évolution des Mondes*, Paris, 1957.

A.E. Scheidegger, *Principles of Geodynamics* (Springer, Berlin, 1958)

A.E. Scheidegger, Recent advances in geodynamics. Earth Sci. Rev. **1**, 133–153 (1966)

A.E. Scheidegger, *Foundations of Geophysics* (Elsevier, Amsterdam, 1976)

P.W. Schmidt, B.J.J. Embleton, A geotectonic paradox: has the Earth expanded? J. Geophys. **49**, 20–25 (1981)

A.J. Schneiderov, The exponential law of gravitation and its effects on seismological and tectonic phenomena: a preliminary exposition. Trans. Am. Geophys. Union **3**, 61–88 (1943)

W. Schröder, H.-J. Treder, Einstein and geophysics: valuable contributions warrant a second look. Eos **78**, 479–480 (1997)

W. Schröder, H.-J. Treder, Geophysics and cosmology—a historical review. Acta Geodaetica et Geophysica Hungarica **42**, 119–137 (2007)

E.L. Schucking, Jordan, Pauli, politics, Brecht, and a variable gravitational constant. Phys. Today **52**(October), 26–31 (1999)

E.L. Schücking, Jürgen Ehlers, in *Einstein's Field Equations and their Physical Implications*, ed. by B. G. Schmidt (Springer, Berlin, 2000), pp. v–vi

S. Schultz, Morgan to receive National Medal of Science. *Princetonian Weekly Bulletin* **93**(8), 1 and 7 (2003). http://theprince.princeton.edu/princetonperiodicals/cgi-bin/princetonperiodicals

M. Schwarzschild, *Structure and Evolution of the Stars* (Princeton University Press, Princeton, 1958)

M. Schwarzschild, R. Howard, R. Härm, Inhomogeneous stellar models. V. A solar model with convective envelope and inhomogeneous interior. Astrophys. J. **125**, 233–241 (1958)

D. Sciama, On the origin of inertia. Mon. Not. R. Astron. Soc. **113**, 34–42 (1953)

C.T. Scrutton, Periodicity in Devonian coral growth. Palaeontology **7**, 552–558 (1965)

G. Shahiv, J.N. Bahcall, The effect of the Brans-Dicke cosmology on solar evolution and neutrino fluxes. Astrophys. J. **155**, 135–143 (1969)

I.I. Shapiro et al., Gravitational constant: Experimental bound on its time variation. Phys. Rev. Lett. **26**, 27–30 (1971)

W.-B. Shen et al., The expanding Earth at present: Evidence from temporal gravity field and space-geodetic data. Ann. Geophys. **54**, 436–453 (2011)

J. Singh, *Great Ideas and Theories of Modern Cosmology* (Dover Publications, New York, 1970)

P.J. Smith, Evidence for Earth expansion? Nature **268**, 200 (1977)

P.J. Smith, The end of the expanding Earth hypothesis? Nature **271**, 301 (1978)

P.J. Smith, An expanding Earth? Nature **278**, 12–13 (1979)

J. Solomon, Gravitation et quanta. Journal de Physique et la Radium **9**, 479–485 (1938)

D. Stanley-Jones, Cosmical zero, and the origin of radiation and dense matter. Nature **164**, 279–280 (1949)

G. Steigman, Particle creation and Dirac's large numbers hypothesis. Nature **261**, 479–480 (1976)

G. Steigman, A crucial test of the Dirac cosmologies. Astrophys. J. **221**, 407–411 (1978)

J. Steiner, The sequence of geological events and the dynamics of the Milky Way galaxy. J. Geol. Soc. Aust. **14**, 99–131 (1967)

J. Steiner, An expanding Earth on the basis of sea-floor spreading and subduction rates. Geology **5**, 313–318 (1977)

A.D. Stewart, Palaeogravity, in *Palaeogeophysics*, ed. by S. Keith Runcorn (Academic Press, London, 1970), pp. 413–434

A.D. Stewart, Quantitative limits to palaeogravity. J. Geol. Soc. Lond. **133**, 281–291 (1977)

A.D. Stewart, Limits to palaeogravity since the late Precambrium. Nature **271**, 153–155 (1978)

A.D. Stewart, Quantitative limits to the palaeoradius of the Earth, in *The Expanding Earth, a Symposium*, ed. by S. Warren Carey (University of Tasmania, Hobart, TAS, 1983), pp. 305–319

G.J. Stoney, On the physical units of nature. Philos. Mag. **11**, 381–390 (1881)

P. Sudiro, The Earth expansion theory and its transition from scientific hypothesis to pseudoscientific belief. Hist. Geo- and Space Sci. **5**, 135–148 (2014)

D. Tarling, M. Tarling, *Continental Drift: A Study of the Earth's Moving Surface* (Bell & Sons, London, 1971)

E. Teller, On the change of physical constants. Phys. Rev. **73**, 801–802 (1948)

E. Teller, Are the constants constant?, in *Cosmology, Fusion, and other Matters: George Gamow Memorial Volume*, ed. by F. Reines (Adam Hilger, London, 1972), pp. 60–66

E. Teller, J. L. Shoolery, *Memoirs. A Twentieth-Century Journey in Science and Politics* (Perseus Publishing, Cambridge, 2001)

P. ten Bruggencate, Die Altersbestimmung von Sternen. Bemerkungen zur Jordanschen Kosmologie. Z. Astrophys. **24**, 48–58 (1948)

D. ter Haar, Cosmogonical problems and stellar energy. Rev. Mod. Phys. **22**, 119–152 (1950)

H. Termier, G. Termier, Global paleogeography and Earth expansion, in *The Application of Modern Physics to the Earth and Planetary Interiors*, ed. by S.K. Runcorn (Wiley Interscience, London, 1969), pp. 87–101

M. Terrall, *The Man Who Flattened the Earth: Maupertuis and the Sciences in the Enlightenment* (The University of Chicago Press, Chicago, 2002)

R. Tomaschitz, Faint young Sun, planetary paleoclimates and varying fundamental constants. Int. J. Theor. Phys. **44**, 195–218 (2005)

S. Toulmin, Historical inference in science: Geology as a model for cosmology. Monist **47**, 142–158 (1962)

K.M. Towe, Crystal structures, the Earth and Dirac's large numbers hypothesis. Nature **257**, 115–116 (1975)

D.C. Tozer, Heat transfer and convection currents. Philos. Trans. R. Soc. A **258**, 252–271 (1965)

E. Tryon, Is the universe a quantum fluctuation? Nature **246**, 396–397 (1973)

E. Tryon, Cosmology and the expanding Earth hypothesis, in *The Expanding Earth, a Symposium*, ed. by S. Warren Carey (University of Tasmania, Hobart, TAS, 1983), pp. 349–358

J.-P. Uzan, The fundamental constants and their variation: observational status and theoretical motivations. Rev. Mod. Phys. **75**, 403–459 (2003)

J.-P. Uzan, R. Lehoucq, *Les Constantes Fondamentales* (Belin, Paris, 2005)

S.I. Van Andel, J. Hospers, A statistical analysis of ancient Earth radii calculated from Palaeomagnetic data. Tectonophysics **6**, 491–496 (1968)

J. Van Diggelen, Is the Earth expanding? Nature **262**, 575–676 (1976)

T.C. Van Flandern, A determination of the rate of change of G. Bull. Am. Astron. Soc. **6**, 206 (1974)

T.C. Van Flandern, A determination of the rate of change of G. Mon. Not. R. Astron. Soc. **170**, 333–342 (1975a)

T.C. Van Flandern, Recent evidence for variations in the value of *G*. Ann. N. Y. Acad. Sci. **262**, 494–495 (1975b)

T.C. Van Flandern, Is gravity getting weaker? Sci. Am. **234**(February), 44–52 (1976)

T.C. Van Flandern, Status of the occultation determination of *G*-dot, in *On the Measurement of Cosmological Variations of the Gravitational Constant*, ed. by L. Halpern (University of Florida Press, Miami, 1978), pp. 21–28

T.C. Van Flandern, Is the gravitational constant changing? Astrophys. J. **248**, 813–816 (1981)

T.C. Van Flandern, *Dark Matter, Missing Planets & New Comets: Paradoxes Resolved, Origins Illuminated* (North Atlantic Books, Berkeley, 1993)

T.C. Van Flandern, Gravity, in *Pushing Gravity: New Perspectives on Le Sage's Theory of Gravitation*, ed. by M.R. Edwards (Apeiron, Montreal, 2002), pp. 93–122

D. Van Hilten, Palæomagnetic indications of an increase in the Earth's radius. Nature **200**, 1277–1279 (1963)

D. Van Hilten, Evaluation of some geotectonic hypotheses by paleomagnetism. Tectonophysics **1**, 3–71 (1964)

D. Van Hilten, The ancient radius of the Earth. Geophys. J. Int. **9**, 279–281 (1965)

K. Vogel, Global models of the expanding Earth, in *Frontiers of Fundamental Physics*, eds. by M. Barone, F. Selleri (Springer, New York, 1992), pp. 281–286

M.A. Ward, On detecting changes in the Earth's radius. Geophys. J. Int. **8**, 217–225 (1963)

A. Wegener, *The Origin of Continents and Oceans* (Dover Publications, New York, 1966)

J.W. Wells, Coral growth and geochronometry. Nature **197**, 948–950 (1963)

J.W. Wells, Paleontological evidence of the rate of the Earth's rotation, in *The Earth-Moon System*, ed. by B.G. Marsden, A.G.W. Cameron (Plenum Press, New York, 1966), pp. 70–81

P.S. Wesson, The position against continental drift. Q. J. R. Astron. Soc. **11**, 312–320 (1970)

P.S. Wesson, Objections to continental drift and plate tectonics. J. Geol. **80**, 185–197 (1972)

P.S. Wesson, The implications for geophysics of modern cosmologies in which *G* is variable. Q. J. R. Astron. Soc. **14**, 9–64 (1973)

P.S. Wesson, *Cosmology and Geophysics* (Adam Hilger, Bristol, 1978)

P.S. Wesson, *Gravity, Particles, and Astrophysics* (Reidel, Dordrecht, 1980)

P.S. Wesson, R.E. Goodson, New pathways in gravitational research. Observatory **101**, 105–108 (1981)

T.M.L. Wigley, Climate and paleoclimate: What we can learn about solar luminosity variations. Sol. Phys. **74**, 435–471 (1981)

C.M. Will, Experimental gravitation from Newton's *Principia* to Einstein's general relativity, in *300 Years of Gravitation*, ed. by S. Hawking, W. Israel (Cambridge University Press, Cambridge, 1987), pp. 80–127

J.G. Williams, S.G. Turyshev, D.H. Boggs, Progress in lunar ranging tests of relativistic gravity. Phys. Rev. Lett. **93**, 261101 (2004)

J.T. Wilson, Geophysics and continental growth. Am. Sci. **47**, 1–24 (1959)

J.T. Wilson, Some consequences of expansion of the Earth. Nature **185**, 880–882 (1960)

J.T. Wilson, Continental drift. Sci. Am. **208**(April), 86–100 (1963a)

J.T. Wilson, A possible origin of the Hawaiian Islands. Can. J. Phys. **51**, 863–870 (1963b)

J.T. Wilson, Static or mobile earth: the current scientific revolution. Proc. Am. Philos. Soc. **112**, 309–320 (1968)

J.T. Wilson, Overdue: another scientific revolution. Nature **265**, 196–197 (1977)

R.M. Wood, *The Dark Side of the Earth* (Allen & Unwin, London, 1985)

R.M. Wood, Is the Earth getting bigger? *New Scientist* **81** (8 February), 387 (1979)

X. Wu et al., Accuracy of the International Terrestrial Reference Frame origin and Earth expansion. Geophys. Res. Lett. **38**, L13304 (2011)

S. Yabushita, The Large-Number Hypothesis and the Earth's expansion. The Moon and the Planets **26**, 135–141 (1982)

S. Yabushita, The Large-Number Hypothesis and the Earth's expansion, II. Earth Moon Planet **31**, 43–47 (1984)

F. Zwicky, On the theory and observation of highly collapsed stars. Phys. Rev. **55**, 726–743 (1939)

Index

© Springer International Publishing Switzerland 2016
H. Kragh, *Varying Gravity*, Science Networks. Historical Studies 54,
DOI 10.1007/978-3-319-24379-5